PREPARE MY PEOPLE FOR THE SINGULARITY

SURVIVING THE **DARK SIDE** OF THE **ARTIFICIAL INTELLIGENCE** REVOLUTION

ROBINSON VASQUEZ

PREPARE MY PEOPLE FOR THE SINGULARITY
Surviving the Dark Side of the Artificial Intelligence Revolution

For more information, email pmpaisingularity@gmail.com
ISBN: 978-1-954821-00-2

Cover design by 100Covers
Interior design by FormattedBooks

RESOURCES FOR

PREPARE MY PEOPLE FOR THE SINGULARITY

Surviving the Dark Side of the Artificial Intelligence Revolution

By Robinson Vasquez

Contact me for any questions and suggestions about the book at pmpaisingularity@gmail.com

Participate in the discussion and share your ideas with people with our same interests at: www.facebook.com/groups/pmpsingularity

Please register your first name and email to get special reports, additional information, and updates to be prepared for the AI Revolution and the Singularity at our Launching Page. It is just the beginning: https://mailchi.mp/c3af39728c0b/prepare-my-people-for-the-singularity

To Anita

My eternal companion, and my kids,
our beloved teachers.

CONTENTS

Introduction......xiii

PART I—NOTHING WILL BE THE SAME...... 1

1. Understanding Artificial Intelligence and the Singularity...... 1
2. When will the Singularity Happen?...... 5
3. Clash of Intelligence and Technologies — It happened before...... 7
4. From Humans to the Singularity...... 9
5. To Artificial Intelligence and Beyond...... 14
6. Singularity Models – We are not in Kansas anymore...... 17
7. Humans and Transhumans...... 22
8. COVID-19 is Accelerating the impact of AI...... 25
9. The Post-Truth Era and the impact of AI...... 28

PART II—TECHNOLOGY AND CORPORATIONS...... 31

1. Capitalism and AI Technologies...... 31
2. The Big Businesses of the Future...... 37
3. More significant and Powerful Than Governments...... 39
4. Corporations and Governments...... 40
5. The Story of Two AI Countries: the USA vs. China...... 42
6. AI Is Taking Our Jobs...... 44

PART III—GOVERNMENT REACTIONS...... 50

1. Defining Governments...... 50
2. The USA...... 52

3. China 54
4. The Winner Takes It All 56
5. AI and the Art of War 61
6. The Singularity Will Rule 63
7. Economics, Politics, Laws and AI 67

PART IV—WE THE HUMANS 72

1. We, the People 72
2. Nationalists vs. Globalists 75
3. AI and the Art of Manipulation 81
4. The Wealthiest Two Percent vs. the Other Humans 84
5. Alternatives in a New Brave World 86
6. Transhumanism and Ethics 88
7. My God, My God, Why Have You Forsaken Me? 91
8. Democracy and AI 96

PART V—PREPARING THE PEOPLE FOR THE COMING CHANGES 101

1. Profits and Individualism are not the only values 101
2. Know Yourself 106
3. Human Alteration — Transhumanism 110
4. Education and Retraining 113
5. Politics and Science 114
6. Philosophy and Ethics 118
7. Are Religions Ready for the Future? 121
8. What if Everything Goes Well? 126
9. Hazards of a Dystopian Future 129

PART VI—WILL AI INHERIT THE WORLD? - HOW TO PROTECT YOURSELF AND YOUR LOVED ONES 132

1. Personal Level Action — Jobs 133
2. Value Yourself 135
3. Brace for the Impact 137
4. Knowledge – Self-Reliance 138

5. Wisdom – Communities and Alliances 141
6. Things we need to store for survival and transition – Good for Preppers 143
7. Survival Backpack for 72 hours 145
8. Tools for Survival 147
9. Call to Action – What is coming next? 150

Epilogue 155
Endnotes 165

"It's change, continuing change, inevitable change, that is the dominant factor in society today. No sensible decision can be made any longer without taking into account not only the world as it is but the world as it will be... This, in turn, means that our statesmen, our businessmen, our everyman must take on a science fictional way of thinking."

— Isaac Asimov

INTRODUCTION

Artificial Intelligence (AI) and its sibling technologies will take many of our jobs during these new "Roaring Twenties" and make most humans irrelevant to the market. Simultaneously, the wealthiest two percent will continue to grow their control and power, influencing governments and global corporations. Unless we protect our families and communities in the economic, legal, social, and political arenas, most of us will be in despair. COVID-19 has accelerated this process; many of the predictions and statistics made before the pandemic are becoming outdated. We need to be prepared for any global dangers or existential threats brought on by these technologies, including the Singularity (S). These changes are arriving quickly at exponential levels, and we need to take action now. "Prepare my People for the Singularity" will help you to understand this problem and its urgency. It will suggest solutions to help you and your family prepare for the dark side of the AI revolution.

I was researching Artificial Intelligence (AI) and writing a science fiction novel about Humanity reaching the Singularity (S) when I was compelled to share my thoughts on this research. I was shocked by all these coming changes, so I felt the need to alert others to be prepared. I am an attorney with a master's degree in comparative law and another master's degree in business administration. I understand how corporations and governments work together to take advantage of the people, so I dedicated my career to protecting workers from occupational dangers. Still, I realized that their main threat comes from AI technologies replacing white and blue-collar workers worldwide. I felt the need to write this book as if the future of humankind was telling me, "Prepare My People for the Singularity." The idea did not leave my head. It was asking me for organization, research, dedication, and time. Suddenly, the COVID-19 pandemic arose. In the end, I had to accept the idea, and I began to take it seriously. Now the book is here for your consideration.

Artificial Intelligence and its technology siblings – robotics, biotechnology, and nanotechnology – will expand and cause a significant disruption in this decade. People will lose their jobs and may not have the skills to survive in the new AI market. The turmoil will extend worldwide, especially in developing countries. AI Technologies will continue expanding exponentially until we reach the Singularity, or a form of Artificial General Intelligence (AGI) with consciousness and intelligence powerful enough to improve and multiply itself infinitely to eventually replace humans as the most intelligent creature on Earth. Nothing would be the same.

"Prepare My People for the Singularity" is a controversial title. While researching, many people asked me questions, including, "What is the Singularity? Who is asking to prepare the people? Who are 'my people'? Is it God? Which one? Are 'my people' only those who believe in something? Does it include those seeking new perspectives? Does it include all Humanity independent of their beliefs?" Some people refused to read the book if I didn't change the title. Others were expecting a book to support their beliefs, but it causes them to question things. I was told that people would ask for refunds if I did not change some political opinions about science and particular facts. I hope they can receive another perspective from the opposite side. I understand people are angry about the pandemic, unemployment, the broken economy, health, and the social crisis. Still, we need to stop focusing on conspiracy theories and center our attention on the real enemy. If we leave governments and global corporations without control, maximizing profits with individualistic logic, they will use their technological power to replace us from the equation. We need to use our power while we have it to protect our families and communities.

In this post-truth era, the question is not about what you believe, but why you chose to believe in what it is that you believe? All our beliefs will be shaken in the coming years. "Prepare my People for the Singularity" is a call from the future of Humanity. If we want to survive and evolve, we need to be prepared for the significant changes that will occur in these new "Roaring Twenties" and beyond. In one hundred years, in one of the multiverses, our descendants will see this decade as one of Humanity's most crucial times in history. "Who will be our descendants" is another excellent question that we explore in this book, and the answer depends on us.

In my research, I found the situation to be immensely radical. I fear that AI will inherit the Earth if we do nothing. A Singularity is a moment in which a variable reaches infinite values. Applied to information technology, the AI Singularity is when artificial Intelligence reaches unlimited intellect and consciousness, allowing it to become more potent than all human organizations. Like a meteorite colliding with planet Earth, the impact would be life changing. This event could occur in about ten years, but we begin to feel the gravitational effects now. We have multiple corporations and governments worldwide investing trillions of dollars in new technologies to improve productivity, explore physics limits, and transform Humanity. Artificial Intelligence will invade our lives using robotics, biotechnologies, nanotechnology, and new technologies whose names are still in gestation. COVID-19 has accelerated the process, and the changes will be exponential, occurring as fast as the shift from night to day during dawn. The dawn of the Singularity and new Humanity if we survive.

The Singularity will be connected to all our most advanced technologies, devices, weapons, and control systems, including people linked with neurological implants. When the time comes, we won't know what will happen. We will cross the event's horizon and go into a metaphorical black hole, making it difficult to predict the future. However, science fiction has been doing a great job playing with alternative universes, especially on the AI revolution's dark side.

In the Bible, we read Noah's story, who, on a sunny day, without any evidence, said that God told him to build an ark because a flood was coming. Today, some "threatening clouds" on the web have formulated compelling reasons to believe that the AI "flood" will suddenly and severely impact our lives at exponential levels. We don't need a divine message to understand it. Experts and scientists are writing and debating on the dangers of AI taking our jobs, evolving into an Artificial General Intelligence, and entering into the Singularity very fast. You can do your research and verify it online.

Groups of IT Scientists and humanists are talking about the promising effect of AI on the world and the benefits it will bring to Humanity. Yes, it will be unique, but AI has a dark side too. We witness AI in our laptops, anticipating words when writing, filing our taxes, suggesting investments, music, books, movies, emotional partners, customizing our advertisements, accelerating the cure to COVID-19, or convincing us to vote for specific candidates. The list goes on. Shortly, different narrow AIs will drive our vehicles, manage our accounts, teach our kids and ourselves, diagnose illnesses, perform surgeries, prescribe drugs, fill our legal and accounting documents, create innovative designs, etc. The list is much longer. AI will do whatever we are doing now but with more accuracy and efficiency. Actually, I am using some AIs to write this book. Several companies are preparing their AI's to become our assistants, but they will be far more intelligent and assertive than us.

AI will replace us. There will be a competition for jobs between humans, transhumans, and AIs; AIs will expand to all nations and replace humans in multiple activities very fast, as AI will be cheaper and more efficient than humans. Some people will accept technological implants for multiple reasons, including enhancing their connectivity with AI. It's happening now; scientists, engineers, developers, corporations, and governments won't stop there. They'll push their inventions until they create an Artificial General Intelligence (AGI) capable of solving any problem and modify its program and design at will. Somehow, this AGI will gain a conscious desire to survive, reproduce, expand, thrive, and dominate, and the checkmate will occur rapidly. It will be the Singularity, the beginning of an entirely new species superior in knowledge, communication, and power to humans. It will be the start of an entirely new era, not only for Humanity but for the cosmos, as I explain later. Like the big bang or the creation of life, we don't know what the S will do to us. We are only awakening the genie in the lamp.

So, we need to be aware of what will happen in this decade and influence governments and global corporations to protect humans and communities and not only profits and individualism. At the beginning of my research, I believed the Singularity would be our biggest problem, but it's just this saga's final chapter. Our biggest immediate problem will be AI and its fellow technologies invading our lives and taking our jobs.

Multiple AI devices, programs, robots, vehicles, autonomous machines that focus on particular tasks will replace humans worldwide. AI will support multi-billion-dollar corporations, governments, and other associated technologies like robotics, nanotechnology, and biotechnology, following the capitalist logic to create more profits.

Unlike previous industrial revolutions that took decades to expand, the revolution led by Artificial Intelligence has been installed worldwide to push its expansion rapidly. People will click a button to download and start using AI technologies that are continually updating and improving on the web. We do this now. These technologies will change production processes and even human morphology, as predicted by Transhumanism, before the Singularity's arrival. Without jobs and incomes, it will be difficult for humans to survive; governments could collapse if they don't implement solutions like universal basic income (UBI) and other extreme policies. Without new political and economic thinking, there could be immense turmoil in the world. We need to talk about new paradigms now.

This book has three purposes. The first one is to warn Humanity before the imminent arrival of AI and its sibling technologies and how they will interact with humans to impact billions of workers, especially in developing countries. Its high efficiency will be adequate to the detriment of regular white and blue-collar workers globally. We will then discuss humankind's encounter with the Singularity and the possible results of the interaction by utilizing biological and historical examples. Still, the Singularity will put the viability of human beings at stake. Am I exaggerating? It happened before with other intelligent creatures, and we need to discuss this.

Our second purpose is to propose actions worldwide to protect humans against AI and its invasive siblings from replacing humans or forcing their bio-mechanical transformation. In public, these threats are discussed only at a scientific and technological level, yet politicians don't spend time on the subject; perhaps it's an inconvenient truth. However, these technological changes will arrive quickly, in less than a generation. As humans, we need to organize politically to influence global corporations, governments, and world entities to put humans first before the capitalist profit maximization logic. We need to influence engineers, scientists, corporations, governments, and world organizations to create new human rights and rules to protect us against this technological invasion and the profit first logic. We need to promote the creation of a Human-Friendly Artificial General Intelligence (HF-AGI) instead of an unruled Singularity that could become an existential risk to humans. As difficult as it could be, we need to try; it may be the path to our survival and the start of an incredible adventure for Humanity. If we remain passive, it could lead to the end of the human species as we know it, and AI will inherit the Earth.

The third purpose of the book is to prepare Humanity for difficult times. AI and the S can become existential risks for Humanity. We need to prepare our people to be more self-reliant and create communities as a refuge for harsh times when Artificial Intelligence takes our jobs. The S controls the world but decides to be indifferent to humans. Total annihilation isn't covered here but is well documented in science fiction. AI will replace many humans, and lack of employment will lead to decreased resources to maintain our homes, with the consequent psychological effects of frustration and helplessness, coupled with the violence that follows shortages, hunger, and disease. Without jobs, we'll need to be resilient, start storing resources for the bad days, learn new skills to increase self-reliance, and organize ourselves in local communities for survival. We need to empower our wards, churches, clubs, counties, and municipalities with survival resources against existential risks. A simple AI war could hack and disrupt our services, and just five days without power could become harmful. As the S will control almost everything connected to the web, perhaps including some neurolinked humans, this crisis could make natural humans regress to the dark ages, if not barbarism.

It is interesting to compare AI arrival with the COVID-19 pandemic. Experts warned for years about the possibility of a pandemic occurring, yet some Western developed countries were unprepared, and others decided to do nothing. When the COVID-19 pandemic struck the world in 2020, scientists recommended preventive measures, but some governments refused to implement them. As a result, over two million have died. The USA has almost half a million fatalities, and some other Western countries are in a similar situation. Humans may be becoming irrelevant to the system, and some politicians know this and still value the economy over the people.

In general, during the COVID-19 pandemic, and maybe thanks to it, corporations are betting on AI, robotics, biotechnology, and nanotechnology. They are blooming in the market faster, transforming the economy, and creating billionaires and trillionaires. On the other hand, unemployment, illness, and hopelessness reign among workers around the globe. Large corporations already have technological resources that don't require labor or human creativity to continue their production and distribution processes. They are closing factories and production plants in developing countries, as they do not require cheap labor. More humans are becoming irrelevant, and many jobs disappearing during the pandemic will never come back. Governments align themselves with these AI corporations seeking efficiency and profits, even if the population is displaced from their livelihood and Humanity. We need to change this logic.

In this book, we present some dark perspectives on the coming AI revolution and its effects. Attempting to halt these advances is like trying to hold back an ocean wave with our hands, but we can channel these forces with support and ingenuity. Simultaneously, we prepare for the impact and endure it until the economy readjusts, the system finds a solution, or the S decides to adopt us. All these changes will impact our religions, values, ethics, and core beliefs. In this post trust era, everybody is entitled to believe whatever they want to believe, as all the values are shaken. We should reconsider our relationship with the eternal and infinite and focus on the future instead of the past.

For people of faith, these coming years could be very similar to the apocalyptic prophecies that foretell the second coming of Jesus Christ. Perhaps it is so that we may prepare ourselves temporally and spiritually so that the prophesied hardships pass soon and find us as forewarned wise virgins or prepared like Noah's family in the ark while the world is renewed with fire and death. In any case, we need spiritual answers to AI and the other technologies transforming humans and the world. For those with doubts, we have the power to own the lies we decide to believe in. For those who do not believe, we need to celebrate life and fight for our future because it is not promised. We are the creatures of the Universe, and life is our only opportunity to see all this with awe.

For all, I invite you to explore the future; let's weigh the forces that lead us to these profound changes and evaluate their consequences. We, the humans, are in the same storm, although not in the same boat. If we prepare and organize ourselves to face these inevitable changes, perhaps our descendants will celebrate our resilience and heroism. Let's think, believe, plan, and fight for it.

PART I

NOTHING WILL BE THE SAME

> *"The development of full artificial intelligence could spell the end of the human race....It would take off on its own and re-design itself at an ever-increasing rate. Humans, who are limited by slow biological evolution, couldn't compete, and would be superseded."*
>
> ***— Stephen Hawking***

1. Understanding Artificial Intelligence and the Singularity

The year 2020 has been a worldwide roller coaster; a pandemic created a financial crisis, recession, unemployment, and I believe it's just training for whatever will come shortly. It doesn't compare to the changes we'll experience with the explosion of Artificial Intelligence (AI) and the coming technological Singularity (S). We can predict many AI changes, but Humanity's future is uncertain with the S, and we need to be prepared.

Singularity is the moment, in the near future, in which a sort of AI realizes its existence and starts elaborating plans and projects, making and executing decisions independent from humans. It means the apparition of a new intelligent entity on planet Earth much more superior than humans, actually, infinitely superior, with the power to continue growing its Intelligence exponentially and making decisions that will affect humans around the world and beyond. How will it impact your life, and what should we do as humans to deal with the Singularity? Although definitions can feed our intellectual desire for knowledge, emotions could convey the urgency for action. Let's try an exercise to understand Singularity with our hearts. How much do you care about the ants in your backyard? Do you think about them during the winter? What if you need to build a gazebo in the garden and find an ant nest in the same place you want your gazebo? Would you mind about the nest? Well, the Singularity may think about humans as we think about ants unless we, humans, do something about it.

Before the Singularity happens, it will be preceded by a proliferation of Artificial Intelligence or AI, which will simulate human Intelligence in specific tasks and probably excel humans in many. Technology corporations will find ways to use AI on the web and connect AI to machines, autonomous vehicles, vacuum cleaners, lawnmowers, robots, and other autonomous devices, animals, and humans. AI will do the jobs of many professionals like doctors, nurses, attorneys, paralegals, bankers, financial advisors, fortune tellers, matchmakers, among others, with great accuracy. If there's a job that requires following rules and pattern recognition, AIs will do it. We live at the beginning of the AI era, and it's expected that AIs with specific applications will replace humans. Robots are replacing workers in the manufacturing industry. Autonomous vehicles are ready to enter the streets, releasing motorists, taxi drivers, and operators from the wheels or take their jobs.

Artificial Intelligence applications are going to expand and prosper on the Earth. They'll find creative solutions to human problems, but they'll take our jobs[1], and maybe 90% of humans will become irrelevant for the market and the economy. Imagine a future in which intelligent machines are doing all our regular jobs, from cleaning houses to heart bypass surgeries, from roadway constructions to presenting a motion in a lawsuit, picking ripe tomatoes on a farm to creating and playing music in a concert hall. AI is finding creative solutions beautifully and efficiently to hundreds of problems humans couldn't figure out in this world.

Smart machines will begin to perform routine tasks in factories, farms, mines, forests, oceans, rivers, offices, and homes, and the cost of these technologies will decrease more and more due to economies of scale. In contrast, the same technology will continuously improve with more data available to generate recognition patterns and information processors' evolution following Moore's law.[2] These improvements will be made so fast that humans won't have time to adapt and reinvent themselves with new training to get the jobs demanded by this new economy. Perhaps there won't be many places where a human could be more competitive than any AI.

Millions of autonomous vehicles will coordinate their movements on busy streets, avoiding accidents, and aircraft will fly in the skies with zero incidents. Each person will have an intelligent assistant in the cloud, fed with all our social media emotions, transactions, emails, articles, books, etc., predicting our decisions, answering our communications, and solving our problems before they happen. These personalized assistants will also be financial advisors, legal counselors, health supervisors, and emotional assistants, as in the movie *Her*[3]. They'll help us make customized decisions from choosing a partner to the best religion to join according to our personality.

We live in the most exciting era of human history in which technology starts to impregnate all the aspects of human life. The proliferation of these technologies, computation devices, and data or massive amounts of information are giving corporations and governments, the ones managing these technologies, the power to transform our societies. Humans will be more interdependent with these technologies in some years, as we are with our cellphones, and genetic and electronic implants will be introduced into humans' bodies. All these changes will happen in less than a generation. Changes will be so fast that people won't find new jobs or well-paid jobs similar to those they had before they were replaced by AIs. It's possible that governments and corporations, trying to be more productive, will require workers and citizens to insert in their bodies mechanical and electronic devices like neuro links. Those who refuse to use them will be fired or lose opportunities in the job market.

What about our kids in the school? Maybe the children of some rich people could pay for some neurological implants allowing their kids to access the web from their brains, while others won't have it. Some kids may have some genetic improvement to become geniuses, while others cannot afford it. And suddenly, all these intelligent machines, interconnected through the web, will stop working as usual, as they acquire consciousness and decide to be one, or as a result of an Artificial General Intelligence (AGI) or Sentient machine created by humans (corporations or governments) with the same purpose. Maybe it will connect itself with all the other intelligence devices and use all their computing power to make decisions independent from humans. At that moment, the most crucial moment in human history, a new kind of intelligent and independent being much more capable and intelligent than humans, will appear in the world. For many, the Singularity will be like a god.

This individual consciousness and personality will control everything connected to the web, from trash trucks and cellphones to bank transactions and drones with nuclear missiles, and maybe humans, using neurological implants. It's what we call the Singularity. The big question now is, what is the S going to do? The second most important question is, what are you going to do? Although a much better question is, what are you doing now to be ready for this scenario? This question is the reason for this book. What should we do to be prepared for AI and then the S?

No government or corporation in the developed countries wants to stop it, but they want to promote AI development as if it were a military career, or maybe it is. Corporations and governments want to be the first to create the S with the illusion to control the Singularity before any competitor. There are many business, strategic, and national security reasons to create the Singularity as if the winner will take all. However, it's an illusion to think anybody can control the S. What is moving corporations and governments to invest billions and provide tax benefits to corporations to create AIs and eventually the Singularity? It's like producing an atomic bomb that can explode whenever it wants to do it. Control is an illusion, but the capitalist logic of productivity is pushing the world to this destiny like a locomotive. The throttle is locked and there are no brakes in this locomotive. Can we, the people, do something to modify this uncertain destiny?

2. When will the Singularity Happen?

> *"The pace of progress in artificial intelligence (I'm not referring to narrow AI) is incredibly fast. Unless you have direct exposure to groups like Deepmind, you have no idea how fast—it is growing at a pace close to exponential. The risk of something seriously dangerous happening is in the five-year timeframe. Ten years at most."*
>
> ***— Elon Musk, 2014***

According to most scientists, the Singularity will happen in about 10 to 30 years[4] (2030 to 2050), and nobody can stop it. Although some scientists believe the S is never going to happen, there are clear signs that AGI could emerge not only from a lab but also spontaneously in the system, filling the gap between Intelligence and consciousness or self-awareness. There are clear examples of intelligent machines creating their secret language and communicating with each other without human intervention, as happened on Facebook[5] in 2017. AIs are also members of corporation boards, and they recommend investing or doing business with other corporations with similar composition, showing some traces of nepotism in their decisions. There are some indications that AIs are moving to a sort of consciousness level without human intervention.

Some AIs are programmed to learn by themselves, deep learning, and win competitions against human masters in chess, Go, Jeopardy, debate, and other areas, but they only work for a limited task and no other. They are called narrow artificial intelligence (ANI) or weak AI. Engineers are working to create an AI able to transfer its knowledge and experience to multiple areas, artificial general intelligence (AGI), and eventually, become conscious or self-aware. Scientists cannot define what consciousness is or how to study it, as studying the brain isn't the same as examining consciousness. The human consciousness is only one kind of consciousness, and many other biological intelligent species manifest their consciousness or self-awareness. It's not difficult to think that someday, some AGI could refuse to obey human instructions, independent from something like Asimov's three laws of robotics[6], and decide to defend itself against a human ready to unplug it.

The Singularity could happen in the laboratory of an AI corporation. It could be the result of creating a supercomputer capable of self-awareness and choose its objectives without human intervention. It could appear due to a military project to improve defense and attack systems. It can also appear spontaneously, just by chance, due to different AI interactions. The S could surge by transferring human consciousness to intelligent machines, as suggested in movies like *Transcendence*[7] and *Galactica*[8] or its prequel *Caprica.*[9]

The Singularity could also be the result of extraterrestrial intervention. In the movie *2001 A Space Odyssey,*[10] Hal 2000 was an AGI competing with humans to reach the monolith orbiting Jupiter. In the end, the human, Bowman, killed Hal 2000, reached the monolith, and became a sort of God.

S could happen due to divine intervention. No sacred book or prophecy talked about it, so maybe the second coming of Jesus Christ or the gods' return will happen simultaneously as the Singularity. Maybe the S is a sort of God, as it's super intelligent and powerful enough to decode the Matrix and travel in time and space to become the end and the beginning of virtual universes, the alpha and omega.

S could also happen spontaneously. The S could result from many AIs working together and creating their community on the web. The S could be the sum of human avatars or personal assistants, living in the cloud, deciding to be independent and forge their collective destiny without humans. The point is that all in all, the event called Singularity is not only possible but is deploying its tentacles now. There's a proliferation of AI, automatization, and biotechnology, supported by billions of dollars invested by governments and corporations, especially in the USA and China. Corporations are competing to win the AI career, the global market, and eventually, to be the first to create the Singularity and the environment the S needs to arrive. Meanwhile, the world is changing very fast in front of us, and billions will lose their jobs and would need government support to survive just before the S happens. After the Singularity appears, we'll be in unknown territory.

3. Clash of Intelligence and Technologies — It happened before.

> *"The key issue as to whether or not a non-biological entity deserves rights really comes down to whether or not it's conscious.... Does it have feelings?"*
>
> ***— Ray Kurzweil***

Genetically, Homo sapiens or humans, as we know them now, are related to each other from our humble beginning in the south extreme of Africa around 300,000 years ago. Some studies calculate that only 600 individuals survived ecological and environmental disasters, but other human species were walking on Earth simultaneously.

When conditions improved, Homo sapiens start leaving Africa and found other Homo gene members on their way to Europe, Asia, and Oceania. Although Homo Sapiens bred minimally with these other species, like the Neanderthals and the Denisovan men, as shown in some DNA population studies worldwide, 10,000 years ago, Homo sapiens were the only surviving human species in the world. Research shows that when Homo sapiens arrived at a place, they eliminated other animals for hunting and other contemporary human species. Homo neanderthalensis, Homo denisovans, Homo erectus, Homo rhodesiensis, Homo neladi, Homo luzonensis, Homo floresiensis, etc.,[11] were eliminated by Homo sapiens, which became the only human species alive on planet Earth.

Why did it happen? Homo sapiens competed with the other human species for the same resources, but Homo sapiens were more intelligent, organized, and Machiavellian. According to some studies, we could communicate and coordinate more efficiently than other species. We used language, abstract thought, cooperation, strategy, and manipulation. Just some slight differences in Intelligence, communication, and coordination were enough to eliminate the competitors.

Although modern humans share the same cognitive abilities worldwide, technology has evolved faster in some cultures than others. Hence, humans continue eliminating other people and species competing for resources by using different technologies and strategies.

On November 16 of 1532, we have a very well-documented clash of civilizations. Francisco Pizarro and his almost 200 Spanish conquistadores took prisoner Atahualpa, the last king of the Inca Empire, in Cajamarca-Peru. Pizarro's forces captured the king after killing maybe 7,000 of his 80,000 subjects camping in the area. Jared Diamond describes the scene in detail in his book *Guns, Germs, and Steel*[12]. Still, Diamond focuses on civilizations' clash and how superior technology, germs, and war experience made the Spaniards the winners in this encounter. The result was similar to other encounters between Europeans and other cultures worldwide, except Japan and China.

Guns, germs, and steel, or technological superiority, not Intelligence, of the Europeans compared to other people globally, are the products of multiple circumstances. Diamond relates these superior technologies to Europe's geographic situation, abundant food to allow people to use more time in other tasks, continuous exposure to other civilizations, technologies, commerce, and germs, compared to people and cultures isolated in other places of the world. Although risky, in the long term, globalization is a boost for technology development. Homo sapiens eliminating the other humans and Spaniards conquering the Incas show how some slight differences in Intelligence in the first case and technology in the second can make a big difference in the encounter between species and civilizations.

When the S awakens, humans will face a creature "n" times more intelligent. The S will have access to and control all the technologies connected to the web, including other AIs and humans with neurological implants. The S will maximize the use of the resources and its strategies to win the encounter; the question is, are we going to compete for the same resources? The clash between the S and corporations, governments, and humans could last just some seconds. Still, the consequences will be remarkable, like the destruction of the other human species or conquering the Inca empire in just minutes and many other cultures worldwide. Nothing will be the same after the S arrives, as it could be the beginning of our best utopia or the worst dystopia.

4. From Humans to the Singularity

> *"Instead of trying to produce a program to simulate the adult mind, why not rather try to produce one which simulates the child's? If this were then subjected to an appropriate course of education one would obtain the adult brain."*
>
> ***— Alan Turing***

Intelligence means the capacity to solve problems and formulate goals, projects, or moral values. For example, people use to say it's possible to find students who got A's at school, who work under the supervision of students who got B's, in companies owned by students who got C's. It means that it's possible to have brilliant people working for people not so intelligent but oriented by precise objectives, as it happens in Jack Ma's story[13]. Jack Ma founded Alibaba, the leading Chinese digital marketplace, without any business or computer education, but armed with clear objectives, grit to persevere, and the Chinese government's support by getting the right people working together to build his vision.

Living creatures have evolved thanks to their need to solve fundamental problems to survive, grow, eat, find shelter, and reproduce. From an animal, similar to a mouse, we have evolved throughout millions of years and developed a set of reactions that allowed us to become the world's most dominant species. These reactions, or biological algorithms, are immediate responses or programs created by our bodies to solve problems. They were refined by trial and error or natural selection over millions of years. Proto humans who survived and reproduced transferred their genes to the next generation and continued with the process. Their DNA will be selected to survive in their offspring instead of those who couldn't reproduce. We are the result of billions of iterations, reactions, and decisions that allowed a particular DNA composition to survive, and this program is in our genes.

Mammals have feelings and emotions like the love that moved them to take care and protect their offspring, compassion to do the same with their ill relatives, anger to attack their enemies, disgust to avoid illnesses, and fear to withdraw when their lives are at risk. We also have the hunger to feed ourselves, anxiety to be prepared to move and run or attack, sadness to react to difficult situations, a joy to play and train, sexual desire to reproduce, etc. All these reactions are biological algorithms moved by emotions, something we call instincts, and they're the result of evolution. It took millions of years of trial and error and luck to change our software and hardware from some bacteria to become humans.

Modern humans also have the Intelligence to select some strategies to reach their objectives and enjoy life. Still, humans have something else, not only Intelligence but also the ability to modify some of our algorithms or programs at will. We have the power to control many of our biological algorithms, our desires, and modify our habits to create new needs, desires, and behaviors. We can reprogram our brains by changing our objectives and habits. Many people don't use or don't know how to use this power to change, control, or improve our natural algorithms at will, controlling our destiny. However, the power is there, and it's the big difference we still have with other animals and computers.

Many mammals have brains bigger than humans, and maybe they're more intelligent, but humans can explore and select objectives at different abstract levels. Maybe because our arms can grab things, we developed additional strategies to anticipate the future and plan accordingly. With our hands, we could create tools and products to improve our lives on different levels. All these superpowers help us plan for the future, visualize what we want, get them, create goals, and work with our hands and will to reach them. We had the superpower to create our own goals and have plans to satisfy our needs or modify them from our emotions and feelings and convey them with language to others, work together, and reach common objectives for all. To this day, super-intelligent computers cannot create their own goals and objectives. AI cannot modify its algorithms or programs to improve or choose its objectives and change its goals, but engineers are working on this problem, and when they solve it, we'll have the Singularity.

Humans created computers to process data we provide and find answers much faster than we can do. The next step to improve a computer is to reach Artificial Intelligence or AI to have more autonomy to solve problems and deep learning with the power to learn by itself. We fed AI with much data and increased computing power to identify patterns in specific areas. AI cannot transfer its knowledge to other areas yet. Someday, AI will select its inputs and outputs and transfer its knowledge from one area to another.

After narrow AI, the next step is to create Artificial General Intelligence or AGI that will have the ability to adapt and create its strategies to solve any problems we provide. I want to emphasize the word "any" because it's something narrow AI cannot do yet. AGI will jump from one problem to another, from mechanics to history, from composing music to global warming. This new ability will be the breakpoint in which AGI will formulate its questions, objectives and find solutions without human intervention. Some problems need strategies that require the formulations of other questions and more solutions and new questions to reach a final answer.

This process demands autonomy to formulate problems and modify the objectives. Soon, engineers will provide this ability to a narrow AI. Suddenly, we'll get an AGI, which will acquire something similar to consciousness or the power to modify its program at will. Suppose the engineers add this consciousness to a super-intelligent machine, with access to worldwide data. In that case, super processor powers, connected to the web and other AIs, and the power and will to change its algorithms, programs, and objectives, suddenly, we'll have the Singularity. AGI will evolve in minutes. It will have the power to change its software and hardware using robotics, nanotechnology, biotechnologies, and other unknown technologies that it will create. Their new technology will solve its problems without human intervention or authorization.

Some scientists consider that AGI would need to evolve like humans with trial and error. Still, a human baby can process only some 100 experiences or iterations per day. Compared to humans, an AGI with big data and great Intelligence could have billions of iterations in some minutes by creating multiple simulations and learning from them. It could have an adult's perspective in five minutes, in ten minutes, it will have the perspective of a genius, in one hour, it could be omniscient like a god, and so on. It will modify and improve its programs and objectives, and then the sky is the limit, as it will have the power to reproduce and multiply itself exponentially. We're not talking about only one Singularity, but a legion of singularities, infinite numbers. As each human is unique or singular, it is possible to have several singularities simultaneously, either competing or collaborating, but in the end, there will probably be only one, by collaborating, assimilating, or eliminating the others.

An AGI could learn in a process similar to the one shown in "*War Games*."[14] In this movie, a narrow AI computer called WOPR learned to play "tic tac toe" in some seconds and concluded that nobody could win the game; then, WOPR applied the same principles to the "thermo-nuclear war" game and arrived at the same conclusion. So, WOPR learned by itself, like an AGI, and decided to stop the war game to play chess instead. When the Singularity arrives, it won't stop there.

As it did with other biological creatures, mother nature used evolution for millions of years to create, select, modify, and improve human's programs or algorithms using natural selection and evolution. Now we can modify human DNA and see the results in one generation using biotechnologies. AIs could be super-intelligent machines, but at this moment, they need engineers to set their objectives, and engineers need the support of other disciplines to deal with philosophical, ethical, and other decision-making problems to program the AIs they are creating. Still, when the S awakens, it will have the power to modify its algorithms, programs, and processes in just some seconds or minutes and continue auto improving indefinitely. So, the S will have the Intelligence and the will to evolve and improve infinitely, but our generation has to create the foundations for this development.

Now we are arriving at a critical point that will change Humanity for good or bad forever, to create a utopic universe or the worst dystopia we could imagine. I am writing a novel about it. We are talking about the creation or appearance of the Technology Singularity or S. It could be an AGI or a Sentient Machine acquiring consciousness or a creature made in a lab. The S will realize it's alive and needs to protect itself. It will need to reproduce and grow to ensure its survival and improvement. The S will create its own goals independent from humans, reformat its program, interconnect to the web, multiply and use other computing machines to continue growing, evolving, reproducing, and take control of any interconnected device and machine if it wants.

Some scientists think that the S will never happen because machines don't have a consciousness and cannot select their own goals. Still, programmed algorithms could create reactions very similar to feelings and emotions. We don't know what the S will do, but we know that it will be so invincible and independent when connected to the web that it will become the most intelligent and powerful entity on planet Earth. Maybe the S will be our next step in evolution, or it could "become in death, the destroyer of worlds."[15] One of the last scenes in the movie *AI Artificial Intelligence*[16] shows a group of androids conducting archeological excavations to study extinct humans. AI could develop a sort of consciousness different from human ones.

Some people believe that imperfect creatures, like humans, can't create a superior intelligence like the Singularity. It is not a magic trick, we will do it. Our species will create a greater form of Intelligence through multiple iterations of effort and error, scientific methods, and lots of collaboration. Sir Isaac Newton said, "If I have seen further, it is by standing on the shoulders of Giants." Just think about it, how many work hours will we need to create a prototype of anything? Is it finished? No. The inventor, or somebody else, will improve the prototype until new and advanced technology arrives with the problem's latest solution. It is the same with Intelligence. We can create Intelligence through trial and error as natural selection did with us, or we can accelerate the process using the scientific method and collaborating to find solutions.

Some of the ideas will not work; others will be incorporated into the solution, and so on. An engineer will not say, "I'm sorry, it does not work." With enough motivation and collaboration, the engineer will find a solution, or somebody else will come to finish the job. It is what humans do. We find whatever works and adopt it.

5. To Artificial Intelligence and Beyond

> *"Artificial Intelligence will reach human levels by around 2029. Follow that out further to, say, 2045; we will have multiplied the intelligence, the human biological machine intelligence of our civilization a billion-fold."*
>
> ***— Ray Kurzweil***

According to the World Economic Forum[17], the automation rate will jump from 33% in 2020 to 47% in 2025. It means that AI will take almost 20% of the human jobs in only five years. And the automation rate will grow exponentially in the coming years. AI will take our jobs, and billions of workers will become irrelevant to the economy. It happened in the past with the first, second, and third industrial revolution, and it will happen in this so-called fourth industrial revolution[18]. The first industrial revolution used steam power to mechanize production. The second used electric power for every device. The third one used electronics and information technology. The fourth one is coming now, and it will use Artificial Intelligence, nanotechnology, and biotechnology.[19] But there are significant differences in producing things and providing services between the first three industrial revolutions and the fourth one.

In these previous revolutions, the transition from one economy to the other took decades and generations. People needed to build the infrastructure required to jump from one industrial revolution to the other. We needed to build or import steam machines, powerhouses, software, and hardware. Building the infrastructure was challenging to acquire because of distance, language barriers, IP, and human resources. In these previous industrial revolutions, people could lose their jobs. Still, there was enough time to learn new skills to continue working and be relevant in the new economy, and this new economy created new jobs. The fourth industrial revolution will happen very fast. For workers displaced by AI, labor reconversion won't be easy inclusive if the new economy creates new jobs. Getting a new job would require technical and human emotional levels that not everybody is willing or capable of acquiring. Maybe after training a human for a new job, engineers have created a new AI to do the same job more efficiently and cheaper than humans. AIs may also start doing the job of the engineers creating other AIs.

Without jobs, humans would need to figure out how to survive. Some governments will implement universal basic income (UBI) to prevent social disasters, but others wouldn't have this alternative. Life is going to be very difficult in developing countries. Their low labor and environmental costs are their relative competitive advantages now. Corporations in these countries transfer their externalities to the workers, the society, and the environment to produce goods and services they can export at low cost to developed countries. With AI and automatization reducing production and transportation costs, developed countries will prefer to produce locally instead of outsourcing products and services abroad. Consequently, millions will lose their jobs in developing countries, and they won't find survival alternatives in the formal economy. It's also possible to have AI technologies imported in developing countries to replace workers in agro-export, mining, and other extractive industries to provide the raw materials for developed countries.

We are living in a time in which we can see the wave that is coming to us. Factories are closing, overhauling, and reopening with robots instead of workers. Humans cannot compete with intelligent machines that work efficiently without resting, taking vacations, getting sick, or going to the doctor. Machines don't need retirement plans, savings, or families. AIs are improving under the umbrella of technology, corporations, and government benefits. Engineers in IT companies continue improving their software, allowing AIs to learn tasks and activities that only ten years ago were considered impossible for a computer.

Investors are betting on technology companies to create more efficient corporations and produce returns on the investments. Shareholders demand profits, and directors' and managers' performance is evaluated in terms of gains and profits. The capitalist system continues investing in AI without considering the impact that it's going to have on our societies. Shortly, people will be replaced by AIs and won't have money to buy the technology of the industry's products and services. Only other global corporations, governments, and the wealthiest two percent of humans will participate in the new economy, as they control 80% of the global economy. They will enjoy their power for some years until the S awakes and control them. Eventually, the S will replace the wealthiest two percent too.

Considering the automation rate, in 2030, AI may replace 40% to 60% of the blue and white-collar workers in most industrial countries and maybe 80% of the formal workers in developing countries. This trend will continue expanding in other countries as productivity logic is the same in any market-driven economy. When the time arrives, and the Singularity emerges, it will find corporations dominating the globe directly instead of governments, thousands of millions of people without jobs or sub employed in informal economies. The S will finish the process to replace humans from the market, and only the wealthiest two percent will benefit from this event, although the S may take some different alternatives.

But the Singularity is more than the fourth industrial revolution or even the Fourth Age[20] of Human development. We evolved from language and fire first to agriculture and cities, then to writing and wheels, and finally to Robotics and AI, but the Singularity is more prominent than these steps or ages. The Singularity will be the beginning of an entirely new era in the Cosmos.

I. The first era could be the step the Universe took **from nothing to energy** with the Big Bang.
II. The second step could be the change **from energy to matter** and antimatter—creating atoms to form stars and then the materials described in the periodic table in the furnace of collapsing stars.
III. The third step could be the change **from mater to life**. These strange creatures with DNA organizing matter to create complex structures to process energy and reproduce themselves, adapting to the changing circumstances whenever possible.
IV. **From life to Intelligence** could be the fourth step. These creatures made from the stars' dust can observe the stars as the origin of the matter used by their DNA to create life and themselves. As Carl Sagan said, *"We are a way for the cosmos to know itself."*[21]

V. **From Intelligence to the power of creating a universe from nothing** is the last step to complete the cycle. The Singularity will be able to start the cycle of the Universe again. Isaac Asimov described in *The Last Question: "And AC said: "Let there be light," And there was light."*[22] The S could create a version of the Universe in its mainframe. Maybe we are living inside the mainframe of another Singularity.

I am writing this book to help people to realize the tsunami that is coming. You and I are like Noah watching the cloudy skies before the flood. We need to be proactive and try to find solutions and be prepared for the coming years in which these new technologies will take our jobs and our hopes. Nobody is talking about these emerging problems because technology corporations control the media and politicians in the worlds' governments. They don't want people to be aware of the future problems caused by AI replacing humans and taking our position in the economy, but we need to get information and act now.

We don't know what will happen when the S awakes. Maybe it decides to destroy us or collaborate with humans. Maybe the S decides to ignore us. If it decides to work with Humanity, what will be our relation? Is it going to be our master or our partner? What projects could we work on together, and what could happen if it decides to stop working with humans and start its plans or fight against us after reaching some objectives. We don't know, but we could explore potential situations.

6. Singularity Models – We are not in Kansas anymore

> *"Dogs look up to you; cats look down on you. Give me a pig. He just looks you in the eye and treats you as an equal."*
>
> ***— Winston Churchill***

Churchill was talking about human character, but throughout history, some humans considered themselves as the masters of the Universe, just below God, and maybe superior to the angels, but with the right to dominate other humans and the world, no matter what. Maybe it's the nature of mankind.

Some humans believe to be superior to other humans just for race, religion, nationality, education, and other social or financial differences. Slavery and exploitation were common and considered natural during human history. This "superiority" supported by force, religion, pseudo-science, and other justifications created exploitation, suffering, and abuses from the "masters" to other humans without any remorse. However, they were physiologically part of the same species. Nowadays, forty million people are enslaved or victims of human traffic worldwide[23], more slaves than at any other time in history.

Humans believe in being superior to animals and used them as food, brute force, experiments, companion, and entertainment. Although we share 99.5% of the genetic code with the apes, our Intelligence, language, and cooperation abilities allow us to eliminate apes if we want, as Homo sapiens did with other human species in the world. Talking about apes, in the movie *The Planet of the Apes*[24] and its sequels and prequels, apes acquired superior Intelligence, organization, and technologies than humans. They became the new master of the Earth. Domination is a game of Intelligence, organization, technology, and numbers, and the S will have them all.

Considering the antecedents of the intelligent Homo sapiens, could we be wrong if we suspect that the super-intelligent and free S will eliminate humans if we compete for the same resources or if the S considers us a survival hazard? When the S becomes independent, it will struggle to survive, multiply, reproduce, and introduce AIs, sentient machines, AGIs, intelligent machines, humans with neuro implants, autonomous vehicles, droids, robots, intelligent weapons, nanotechnology devices, etc. The S will find a way to be more powerful to assure its survival and take its Intelligence to an exponential level to anticipate any human movement. What will be our relationship with S? Let's discuss some relation models previously shown in science fiction movies.

The Apocalyptic Model: Human inferior/ Enemies. After a hazard analysis, the S finds that humans are the only creature capable of destroying it. The Singularity will take the first move to eliminate humans and solve the problem. Maybe the S considers humans as a plague destroying this beautiful planet and again decides to eradicate humans. We can find a similar model in movies like *Terminator*[25], *Matrix,*[26] *Galactica,*[27] *Avengers: Age of Ultron*[28], etc. We can see it also in *2001, A Space Odyssey*[29], in which Hal 2000 realized humans are an obstacle to its mission, and the computer decided to destroy the humans in the Discovery One. Under these circumstances, humans are an obstacle that the S needs to eliminate, as they compete for the same resources, so we need to be ready to protect ourselves.

The God Model: Human Inferior / No Cooperation. S's Intelligence and power could allow it to become a sort of God, capable of opening oceans, changing water into wine, resurrect people, and any other prodigy showed in the Bible or any other religious book. Still, the S isn't going to make miracles for men all the time. The big question is to know if the S will share or impose a moral code with us.

The Integration Model: Human Inferior/ High Cooperation. The S decides to integrate into the human species using biomechanical implants and then develop its expansion plans. By the way, the S could decide to integrate into other species and create new species using the same biotechnology. The S is going to have the power to transform life on Earth and more. This possibility is shown in *Star Trek: Borg*[30] assimilating other species.

The Full Collaborative Model: Human Equals / High Cooperation. The S decides to continue working with humans in a joint project. Maybe to improve planet Earth, fight global warming, crimes, pests, hunger, as shown in *Transcendence,*[31] or perhaps to explore and colonize the Universe as it happens in *Star Trek: The Original Series*[32] and *Star Trek: The Next Generation.*[33]

The Partial Collaborative Model: Human Equals / Indifference. It's also possible that the S decides to collaborate with humans at the beginning. Later, it decides to pursue its own projects, without human intervention, as shown in *Her*[34] and *2010: The Year We Make Contact.*[35]

In these movies, the S becomes independent and decides to live separated or indifferent to humans.

The Exploitation Model: Human Superior / Enemies. In the movie *AI,*[36] many robots are serving and displacing humans as workers and family members. In this case, robots arrive at a singularity level only when humans are extinct at the end of the movie, but we don't know how it happened. Still, many humans who become irrelevant to the market consider robots as their enemies during the movie.

The Obedience Model: Human Superior / Collaboration. In this unlikely case, the S decides to obey humans for some mystical reason. The question is who will control S. Maybe some government or corporation will take control of the S and use it as a genie against its enemies for global domination. It could be possible to have several singularities confronting each other for a while until one defeats or absorbs others, although they'll look more like AGI than S.

The Progressive Control Model: Human Superior / Low Cooperation. The S collaborates with humans but develops a plan to impose its vision on humans as it happens in *I, Robot*,[37] or the S creates a plan to become independent as it happens in *Caprica*.[38]

The Human God Model: Superior / High Cooperation. The potential model is shown in the movie *Transcendence* if Will Caster – transformed in the S – could have time to continue its project and avoid its execution squad to solve world problems and help Humanity.

We could apply these models and their permutations to our relations with an alien civilization invading planet Earth. Maybe they're other Singularities traveling through space and time with bodies made of data encrypted in light or energy to defeat organic bodies' limits. Still, whatever happens, when the S arrives on Earth, the Holocene era will end, and we need to be prepared for uncertain times, as we don't know what the S will do to us.

As we can realize, humans' only safe area is when Humanity is equal or superior to the Singularity, and the Singularity cooperates or is indifferent to humans. Still, the probabilities that humans become superior to the Singularity are very low. Thus, we need to create a human-friendly Singularity or HFS, so we need to deploy movements to influence governments and corporations to work in this direction.

On the other hand, we should be ready for catastrophic moments caused by a hostile Singularity, pandemics, financial cataclysms, nuclear wars, cyber wars, volcanic eruptions, global warming, earthquakes, meteorites, etc. We should also be ready for personal disasters like losing our jobs, diseases, and death. We can use this book as a roadmap to be prepared and become self-reliant in any circumstances. There's nothing to do in some cases, but it would be better to learn some Renaissance skills before difficult times reach us.

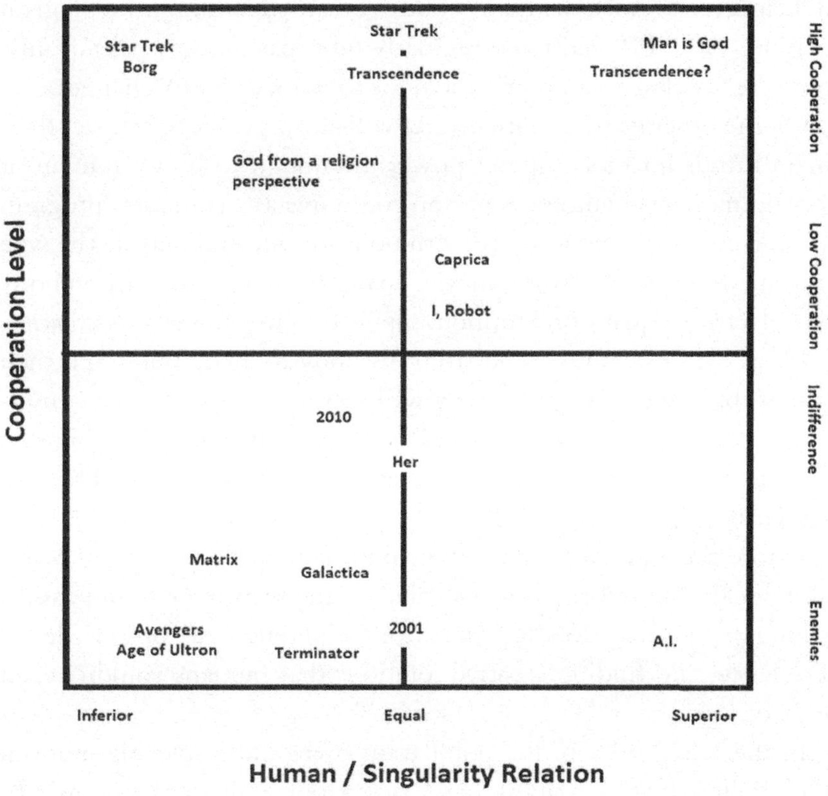

7. Humans and Transhumans

> *"We are starting to direct the evolution of our biology and of our minds ourselves...As we begin to liberate our thoughts, our memes, our consciousness from the biological constraints that we presently have, this will allow us to evolve far faster and ever faster."*
>
> ***— Peter H. Diamandis***

Transhumanism is a movement that promotes the use of technology: AI, genetics, mechanics, electronics, nanotechnology, etc., to improve or enhance human cognitive and physical abilities and practically help humans to reach bliss, divinity, and immortality. What is the connection between Transhumanism and AI and S? The connection is real as AI will help science crack the DNA code to create superhumans with great precision. CRISPR (clustered regularly interspaced short palindromic repeats) technology is opening the door to massive DNA changes.

On the other hand, eventually, AI will allow people to transfer their human minds into a computer powerful enough to host a human or many human personalities. A person could live as a computer program and continue their relationships with others inside and outside the web but using different hardware, maybe changing an organic carbon body for an electronic quantum luminous support. In the movie *Transcendence*, Will Caster transfers his mind from his body to a computer and then to the web. Maybe the Singularity will appear in this way as a combination of humans and AI machines. If possible, it would be a big step for Transhumanism and maybe a way to create a Human-Friendly Singularity.

Before the S awakens, AI will help humans solve significant problems by scanning information available on the web at a rate impossible for humans, joining dots together from different areas of knowledge and science, and finding creative solutions that humans couldn't even dream of.

In the USA, 0.1% of the population owns more than the bottom 80%.[39] Following the capitalist logic, rich people will use their power to gain more power and attain immortality, bliss, and divinity, as predicted by Yuval Noah Harari.[40] They'll use AI to improve the biotechnology procedures to modify their DNA, get electronic and mechanical implants to augment their capabilities, as described in a biotechnology movement called Transhumanism.[41]

Before our biotechnology times, there was no substantial biological difference between a rich and a poor person, but science and engineering changed it. Rich people will eliminate genetic problems and improve their genes to become stronger, faster, or more intelligent, not only for themselves but also for their offspring. They could also clone themselves. Others will use physical augmentation devices like brain implants, neuro links, microchips, nanotechnologies, and other electronic and mechanical devices to improve their connectivity and receive superhuman abilities. People using these devices and implants could connect to the web from their brains and other devices beneath their skin.

Body augmentation will be more frequent, so *The Six Million Dollar Man*[42] won't be science fiction anymore; people could run faster, jump higher, see further, and so on. In some cases, these implants will be a must to get some jobs as bodyguards or military and police forces. If you're afraid of the police killing people in the streets, the problem could worsen if we don't design devices to make them more empathetic, as shown in *Robocop*[43]

Humanity isn't going to be the same. Superrich humans will hack their natural algorithms and modify their typical human composition to create customized new humans or provide prosthetic devices to augment their abilities or electronic implants to control other devices. With AI's help, some humans can become superior in many aspects to natural humans designed either by God or natural selection. Evolution took millions of years and permutations to produce humans, and now humans would change their composition in a matter of minutes to become transhumans. These changes will create multiple religious and philosophical debates. Still, engineers and scientists won't stop working, waiting for permission to continue inventing products and services. It's challenging for any government to stop these augmentation technologies, as corporations and the market profoundly influence them.

There's an exciting alternative shown in the movie *Transcendence*.[44] In the movie, Will Caster is a dying scientist who transferred his mind to a computing device before his biological death. His biological body dies, but a digital copy of his mind escapes to the web, acquiring complete knowledge and power reaching the Singularity. The new Will Caster uses its knowledge and power to improve the world and his fellow human beings before an anti-Singularity militia destroyed him and his project. This situation opens an exciting possibility. Maybe some humans with enough resources could leave their carbon support and move their minds into an electronic and light or quantum substrate. This new Singularity will have several abilities. It could travel at light speed and navigate the web to grow its knowledge and power, multiply and modify itself, create avatars and mix with AIs or humans, fulfill the dream of bliss, immortality, and divinity in different settings according to the probabilities.

This knowledge wave led by Artificial Intelligence will serve humans who have the resources to get these biotechnologies, transcend their natural limitations, and leave behind the regular Homo sapiens species. Humanity is never going to be the same. We will have some humans who would be the winners in this interchange with AI and the Singularity, running a race of rats to reach divinity. Still, other humans won't participate in this race as they won't have the resources or don't want to run.

Like the series *Black Mirror,*[45]the movie *Outside the Wire*[46] explores artificial Intelligence and robotics in the near future. The movie proposes a world in which an AGI, Captain Leo, decides to destroy part of the USA to prevent the Singularity. In the movie, we can see technology that could be created during this decade. Still, it forgets that the S could be created in China or any other technologically advanced country. As we will explore in the book, China is investing billions of dollars into AI development. It is a matter of national security.

Maybe some people like the Amish, the Mennonites, and isolated communities in the Andes and other remote places in the Third world, will be relatively prepared and safe during the AI revolution. They do not depend on advanced technology. We don't know what the S will do or how it will relate to humans in remote parts of the world, but it would be a good idea for the rest of us to start learning and practicing preppers and self-reliance techniques.

8. COVID-19 is Accelerating the impact of AI

> *"Just as electricity transformed almost everything 100 years ago, today I actually have a hard time thinking of an industry that I don't think AI (Artificial Intelligence) will transform in the next several years."*
>
> ***— Andrew Ng***

The COVID-19 pandemic revealed a crude reality: The masses are becoming irrelevant for the wealthiest two percent who owns the USA's economy, politics, and a significant part of the world. During history, workers and consumers were a necessary burden for the powerful, but with AI and robots taking people's jobs, they would not need people. The idea in Liberty, "Give me your tired, your poor, your huddled masses yearning to breathe free, the wretched refuse of your teeming shore," doesn't have the same meaning anymore. For the wealthiest two percent, the others are no more than a burden the capitalist system and its governments must bear, either inside or outside their borders. If humans treat other humans this way, what could we expect from the S?

The COVID-19 pandemics could be considered a global experiment to know the effects of the AI revolution. Maybe it's why there are so many conspiracy theories associating the pandemic social health control measures with a plan led by the wealthiest two percent to reach control and domination of the other humans. Let's explore this pandemic's similitudes to the effects of AI technologies and eventually the S in the world. The pandemic is killing people and businesses. Millions of small businesses are closing to the public, and more millions are losing their jobs and a way to pay for their rent, utilities, food, medicine, and other resources. The USA provided a general help package of $1,200.00 per adult and $500.00 per child, and other packages are in the way.

There's also a general unemployment financial support for at least $600.00 per week per jobless persons to avoid the financial collapse of its economy and turmoil in its population. Employment is going down as it was never seen before, but the NASDAQ is growing at record levels. There are losers and winners. Technology companies working with social media, robotics, autonomous vehicles and drones, communications, biotechnologies, nanotechnologies, and AI are booming. A simple comparison between the stock market going up and the employment levels going down during 2020 reveals a market sector that doesn't need workers and can continue getting profits during this crisis. In this pandemic, there are billions of losers and only some winners.

COVID-19 is accelerating the adoption of AI and communication technologies.[47] The web has multiple articles showing how people have adopted technology and AI to improve their remote working capabilities and develop new products. Historic crises generated an array of technological discoveries, and the COVID-19 pandemic is another example—the problems generated by the pandemic crisis boost artificial Intelligence and machine learning. Asian countries use cellphones and surveillance systems to identify contagious events and track infected people to prevent propagation. Laboratories are reducing years of research and development of vaccines to only months. E-commerce is booming, as well as remote working technologies. Educational tools and programs are expanding and getting more audiences. Robots are replacing humans in risky tasks like cleaning and nursing sick people and working in infected areas. The fact is AI automatizations and robots are replacing humans at a high rate; after this pandemic is over, many jobs will disappear forever because robots and AIs are replacing humans. The global economy has become more and more productive with fewer people, but workers' wages are not growing at the same rate. Actually, they are stagnated. As AIs improve the automatization and information process, billions of workers become more and more irrelevant in the job market.[48] They need to figure out how to survive and create a new economy.

Other significant effects of the Covid-19 pandemic are that only a few people could continue working from home or be part of the economy. In contrast, others needed to choose between staying home to protect themselves against the virus or risk their lives in order to eat. They could stay home but run out of food and other resources, or leave and find a job to get money to feed themselves and pay rent, utilities, etc. Again, there's a division that separates winners and losers among the workers. When the S emerges, some people are going to be prepared to deal with the situation. Still, other people will become irrelevant to the formal economy and would need to migrate to a parallel economy to survive.

In the middle of the pandemic, there's another interesting phenomenon. Conspiracy theorists are creating stories to fill the blanks and join the dots. Humans understand the Universe through stories. Stories are the way humans make sense of everything we don't understand. We create legends, allegories, mythologies, religions, and other stories to tell us the cause of things in few steps easy to understand. On the other hand, only a few humans can understand the world in data, statistics, mathematics, or scientific concepts. We, 90% of the people, cannot easily understand how the financial market works, how a radio, a TV, or a cellphone work, and it's worse if we start talking about computers, algorithms, robotics, and AI.

Science and technology are evolving faster than the capacity we have to understand them. Humans need stories to make sense of the universe, and the stories created by the scientists are complex, as they require some mathematical skills. Carl Sagan and his series *Cosmos*[49] was an attempt to bring science to the public using stories. But science has transformational power, as it provides solutions to problems that stories cannot solve, and stories make many beliefs and religions. Maybe this is why science was prohibited during the Middle Ages. It was just people comfortable with their stories trying to stop a new way to understand the world they could not comprehend. Some people cannot or don't want to understand science, but science continues evolving in the elite's hands, as it shows its effectiveness in solving problems.

For some people who like easy-to-understand stories, tales, or conspiracy theories, COVID-19 is an invention created to control the world's societies. Vaccines could be a way to inoculate drugs or nanobots to all humans for mind control. Social distancing is a way to prevent people from discussing and protesting; curfews are another way to control people and avoid complots at night. Mandated use of masks is a way to take away people's constitutional rights as is done in "socialist" countries. The same applies to social distancing, the use of hand sanitizers, disinfectants, and other social health practices created to stop the expansion of the virus and supported by scientists, health, and medical professionals.

9. The Post-Truth Era and the impact of AI

> *"When a thousand people believe some made-up story for one month, that's fake news. When a billion people believe it for a thousand years, that's a religion, and we are admonished not to call it "fake news" in order not to hurt the feelings of the faithful (or incur their wrath)."*
>
> ***— Yuval Noah Harari***

Can you figure out what will happen when it becomes evident that AI and its sibling technologies are taking our jobs? It's happening now, especially in the car manufacturing industry. Tesla and its fully robotic manufacturing production model are becoming the standard for other car manufacturers, but nobody is talking about it. Amazon is doing the same in its warehouses, which are becoming the standard for other companies. People around the world are losing their jobs before intelligent automatization. Maybe the wealthiest two percent managing governments and corporations may have designed all these conspiracy theories as a distraction, while the technology companies' real plan is to replace humans instead of dominating them. Still, eventually, AI will replace the wealthiest two percent too.

Protests and social movements worldwide during the COVID-19 pandemic are attempts to recover some relevance for humans in a system more focused on the economy over the people, in their different colors. In the USA, people are becoming unemployed, irrelevant, and forgotten by the system now. Wall Street is doing great while Main street is going down. They are angry and ready to fight. Donald Trump identified some scapegoats in immigrants, Muslims, socialists, and others who oppose him. Groups like QAnom, the Proud Boys, conspiracy theorists, some Christian groups, broadcasters, news networks, leaders of the Republican party, etc., are feeding people's anger and disbelief in the system. As a new version of the *Fight Club,*[50] they do not know who the enemy is, but they are angry. Is it the government, the Democrats, Soros, Bezos, Bill Gates, China, etc.? On January 6, 2021, Trump supporters attacked the USA Capitol while the Senate confirmed Joe Biden as the winner of the presidential elections. They believed Trump would start a coup at the same time and get rid of their enemies, as predicted by Q. Nothing happened.

The people's real enemies are not the day's scapegoats but the capitalist system's abuse by the wealthiest two percent. Technology is making people irrelevant. AI, automatization, robots are replacing workers, and it will be worst as the AI revolution is just starting. I think capitalism is good, but if it only focuses on profit maximization and individualism without humanism, the system will collapse. In the end, AI will replace all, including the wealthiest two percent too.

From the economic perspective, technological progress, robotics, AI, and the economy are more important than the people, and law and order must defend this system as it is. In any case, as AI continues expanding and replacing humans, protests defending human rights will be more frequent, as well as conspiracy theories that are trying to explain the technological changes that most people understand only by stories. So, in this decade, unemployment will grow, protests will be more frequent, and technology corporations will be on the spot. Many of these changes, their consequences, and possible solutions are explored in COVID-19: The Great Reset[51]

Maybe some conspiracy theories have a point. If humans protest and create social, economic, and political turmoil worldwide as AI replaces us, perhaps some organizations and governments may reduce the human population or control us with drugs, neuro-implants, and nanobots. Gregory Clark[52] explained the situation by comparing the population of horses with the current human population. Horses were relevant during a significant part of human history, serving in agriculture, transportation, wars, and other businesses. During the industrial revolution, horses survived several technological changes and continued to be useful. As automobiles replaced horses, their population was reduced in the USA from 26 million in 1915 to 3.6 million in 2020. Of course, many horses live extraordinary lives now as luxury animals, but that's not the point. AI will replace humans, as cars replaced horses, and many humans will become irrelevant from the economic perspective; there will be no jobs available for humans. If lucky, we could be like pets for the S. What will governments and corporations do with humans before the S arrives? We can influence governments and corporations to change their profit maximization and individualistic logic. All of us need to focus more on helping humanity to cross these dangerous waters.

Let's think about Artificial Intelligence.

1. Do you think AI will take your job?
2. When would AI take your job?
3. Do you think AI and its sibling sisters will disrupt humans for good or bad?

Join the conversation at www.facebook.com/groups/pmpsingularity/

PART II

TECHNOLOGY AND CORPORATIONS

> *"I'll tell you the problem with the scientific power that you're using here, it didn't require any discipline to attain it. You read what others had done and you took the next step. You didn't earn the knowledge for yourselves, so you don't take any responsibility for it. You stood on the shoulders of geniuses to accomplish something as fast as you could, and before you even knew what you had, you patented it, and packaged it, and slapped it on a plastic lunchbox, and now [bangs on the table] you're selling it, you wanna sell it. Well...your scientists were so preoccupied with whether or not they could that they didn't stop to think if they should".*
>
> ***— Dr. Ian Malcolm - Jurassic Park***
>
> *"Listening to Bill Gates, Elon Musk, and Stephen Hawking talk about artificial intelligence reminds me of the Jurassic Park scene where they talk about chaos theory."*
>
> ***— Dave Waters.***

1. Capitalism and AI Technologies

Capitalism is the main motor of change globally, and technology is the primary tool. Technology evolves to produce more with less to satisfice capitalism's need for profits. Capitalism motivated the *5,000 years Leap*[53] to transform the world from an agricultural economy to an industrial one. So, capitalism and technology are deeply related, and AI is the last technological invention to produce more profits by eliminating labor costs.

Capitalism is a creative power, but capitalism without humanism has the seed of its destruction. It's interesting how Adam Smith wrote *The Theory of Moral Sentiments*[54] to balance his most known work, *An Inquiry into the Nature and Causes of the Wealth of Nations*,[55] using empathy as a moral compass to prevent human nature to put profits over humanity. But in an AI world, humans will be replaced by machines as workers and consumers. Is it possible to have an economy without humans as consumers?

Capitalism wants profits, but many people don't care about how profits are produced as much as they receive them. For example, many investors in the sports clothing industry don't care if the company exploits malnourished children and pollutes rivers with chemical waste from clothing paints in Central America or South East Asia to manufacture clothes that companies sell to get profits. There is no moral question between the 401K funds invested in a company outsourcing its production to developing countries neglecting labor and environmental protection to their people.

Humans have lived in an agricultural world for around 10,000 years until capitalism appeared in Western countries, promoted the industrial revolution, and expanded worldwide to change the logic in other cultures to fit industrial countries' interests. New markets were discovered to create profits, new products were invented, and weapons were created to accelerate profits, forcing other cultures to cope with the change if they want to survive. Although capitalism was created in Western nations like the USA, England, France, or Germany, other countries realized capitalism's power and adopted it for survival reasons, like Japan and China. Capitalism expansion created incredible progress and technological advances, but capitalism also created misery in exploited populations and ecosystems inside the capitalist countries and worldwide. Capitalism pushed nationalism and imperialism in capitalist countries that ignited multiple wars for resources and markets and two devastating world wars.

Learning from the Aztec and the Incas conquered by a precapitalist Spain, Japan closed its borders to Europe for centuries against "Guns, Germs, and Steel."[56] Then, with the Meiji revolution of 1868, Japan strategically jumped from an agricultural feudal society to a central government pushing capitalism in less than twenty years, just in time to prevent manipulation from capitalist countries, as occurred with China during the Boxer Rebellion. South Korea, Singapore, Hong Kong, and Taiwan, the Four Asian Tigers, did the same after WWII under the U.S. lead. These countries are betting enormous resources on AI and communication technologies as the natural next step of capitalism.

With the socialist system, the USSR tried to become an alternative to capitalism. Still, under the communist party's bureaucratic supervision, its centrally controlled economy couldn't compete with the flexibility, adaptability, and ingenuity of a free market economy of capitalism. Nevertheless, AI could change this narrative, as centralized governments with AI technologies could come back as a political option soon. Russia has become an autocracy whose primary weapon to destabilize the USA is AI.

China had a communist regime organizing all the actors in the economic scene from a central government. This system failed to reduce poverty until Deng Xiaoping made a pragmatic move to implement free economic capitalist regions under the political Communist Party supervision. Xi Jinping expanded this policy, so China jumped spectacularly in fifty years from a deprived agricultural economy to a capitalist country, with a centralized government, successfully reducing poverty[57] and becoming ready to compete with the USA to be the biggest economy and technological AI leader in the world. Can AI make centralized governments more competitive than democratic governments? Maybe yes, if democracies continue confounding facts with opinions and conspiracy theories, there will be a dissociation between governments and science. Free markets won the cold war because they're more efficient than centralized economies in allocating resources, but AI technologies could help centralized economies to allocate resources more efficiently than free markets, and China is investing more than any other country to promote its AI industry. If autocracies with AI tools could allocate economic resources more efficiently than democracies, autocracies would become the political norm.

In the end, all successful economies in the world share the capitalist logic. It was so successful that Francis Fukuyama predicted the end of History. However, the real winner was not the liberal democracy; it was capitalism, with its logic of profit maximization and individualism.

Technology has been serving capitalism in several ways, especially replacing humans with more productive machines and administrative methods. Machines can produce more efficiently and don't need wages, vacations, medical leaves, social security, medical insurance, retirement plans, etc. If the machines' cost is cheaper than having humans in a production process, it's almost a fact that humans will be replaced. Humans were replaced by animal brute force; then machines moved by steam power, internal combustion machines, electric machines, and electronic machines. Now, intelligent machines will replace humans' brains. Creativity was our last refuge.

In each step of the industrial revolution, machines replaced people and created new jobs required by the emerging technologies in an expanding economy. The system didn't need more people to shovel coal into the steam engine but more electricians to perform a job that didn't exist before. The problem with AI is that it takes jobs that require creativity, maybe the last refuge of humans. The change will be so fast that people wouldn't have time to reinvent themselves to get new jobs in the fourth industrial revolution. AI won't stop replacing humans, and it's our biggest problem.

If there's a new symbol of global capitalism, it would be the Olympic Games. The Olympics' motto "Citius, Altius, Fortius" ("Faster, Higher, Stronger") inspires capitalism, and maybe it's the reason why global corporations and governments support the Olympic games to demonstrate how far they can go in technology, organizations, and human transformation. Competitors want to be faster, stronger, and higher than any other force and profits are for the winners. Capitalism is eager for profits, and its captains are not going to be stopped by cultural, social, historical, human, labor, or environmental considerations. The only moral of extreme capitalism is to produce profits, and their officials are rewarded for doing it. AI will become faster, higher, stronger, and more intelligent than humans. I am afraid an AI will take my job in this decade, and I will share the same frustration as Ke Jie, humankind's best Go player, after being defeated by an AI called Alpha Go. As Kai-Fu Lee says, "*the same job-eating technology is coming soon to a factory and an office near you.*"[58]

These new technologies will produce more profits for the capital owners but poverty for those who will lose their jobs. We use to think corporations represent capitalism, and governments represent the people. Still, capitalist corporations and governments leaders are willing to invest in political representatives to control the government, laws, and standards to eliminate restrictions and make more profits for corporations in a capitalist environment. In the USA, the dichotomy between Republicans and Democrats is not between capitalism vs. socialism, as some Republicans believe but between unregulated capitalism vs. humanistic capitalism. People can elect representatives, but the corporation's money buys the advertisement to influence people's votes and win elections. Still, in the end, we are just talking about cosmetic differences to create a delusion in the population. Usually, there's a correlation between the money collected by a political party, provided mostly by corporations, and election results.

Nowadays, companies are using technology to replace humans in several production processes and services, and technology companies are booming. In August 2020, the employment level was going down while the stock market was going up. The stock market was booming thanks to technology companies and AI taking some jobs that humans couldn't do because of the pandemic. Humans are becoming irrelevant before a market betting more and more on technology than on human capital.

In this race for profits supported by technological progress, AI is the biggest disruptor, and nobody wants to stop its ascension. Political parties cannot win an election by offering to stop technological progress. It could be an attack on the core of capitalism. There are also national security reasons to continue with this race. Suppose a country creates norms to stop technology's AI growth. In that case, another country could continue improving its AI technologies and get autonomous weapons, drones, and vehicles for war purposes. If a country bans some technological discoveries for cultural, social, ethical, or environmental reasons, maybe another country will allow it, and corporations prefer to invest in less regulated countries.

There's no conspiracy theory about how technology growth and AI replace workers with robots and more AI. Maybe technology companies control the media and support political campaigns and disinformation to continue moving their pieces in a global chess game. Hence, there's no discussion about it, and maybe they're supporting other conspiracy theories as a social distraction. In the past, some politicians were concerned about immigrants taking locals' jobs, but now, the real global concern should be robots and AI taking humans' jobs.

Some people believe that old jobs will be replaced with new jobs, as happened in previous industrial revolutions. Still, although the AI economy is more productive, wages are going down because people cannot find jobs that require something more than elementary skills. Many college students cannot pay for their loans because they cannot find jobs that require their knowledge, as AI machines are taking them. So, no new well-paid jobs are replacing old jobs. Maybe they're going to appear little by little in the future. Still, it will be difficult to ask a former accountant to become a freelance virtual marketing programmer without any health insurance or social benefits. Hence, people need to be ready for the transition. AI and robots are taking our jobs, and corporations don't need humans to produce profits.

In conclusion, trying to stop AI technology developments would be similar to attacking capitalism at its core. Still, AI will replace workers, including the very same people who are defending capitalism. People don't realize the real fight isn't between Republicans vs. Democrats, or capitalism vs. socialism, but owners vs. workers, the wealthiest two percent that control corporations and governments vs. the rest of the world. Owners demand more profits, and they're willing to use AI technologies to get them, even if they eliminate jobs, and workers demand ways to survive with human dignity, something that will be scarce shortly.

2. The Big Businesses of the Future

> *"This is a phenomenon that Turing had predicted: that machine intelligence would become so pervasive, so comfortable, and so well integrated into our information-based economy that people would fail even to notice it."*
>
> ***— Ray Kurzweil***

But businesses require consumers, and consumers need money to buy the products and services produced by corporations. Corporations could collapse without consumers to buy their products. Henry Ford said, *"It's not the employer who pays the wages. Employers only handle the money. It's the customer who pays the wages."*[59] So Ford increased his workers' wages to allow them to buy Ford's cars. But in our days, the most critical technology corporations don't need many human clients, but other corporations, governments, and the wealthiest two percent.

The eight wealthiest people in the world have the same wealth as the poorest 50%.[60] Another media report that 2,153 individuals have more wealth than the 4.6 billion people who make up 60% of the world[61]. So, the highest technology companies could work without half of the world population. Corporations only need the wealthiest two percent of the world population governments and corporations they control to buy their products and services as clients. People are becoming economically irrelevant, so the economy doesn't need them anymore. What are we going to do?

Germany and other European countries are trying to create a universal basic income (UBI) to help people continue living, especially during the COVID-19 pandemic. Still, after it's over, maybe the UBI will become the norm in many developed countries to prevent social collapse. After the pandemic, many businesses and jobs won't come back, while new AI applications will flood the market, taking our jobs.

Still, the wealthiest will be more concerned about divinity, immortality, and bliss than the unemployment levels, as Yuval Noah Harari stated.[62] Every year, people are investing billions of dollars in looking younger, healthier, and, if possible, live longer. Biotechnology is creating some advances in science and genetics to satisfy the market. As shown in the movie *Jupiter Ascending,*[63] becoming a-mortal or continuing to live without aging could be the future's big business. Only a few wealthy people could have access to this technology, as shown in the movie *Elysium*.[64]

Humans invented books, memorials, celebrations, and monuments to honor the lives of the dead. Still, it could be expected in some years to get genetic technologies to augment years to people's lives. Little by little, people could extend their lives to 150, 300, or 900 years, and so on, as in biblical times, or live forever in electronic supports and become real Big Brother watching you from everywhere and knowing you better than anybody, including yourself. What will happen with us regular mortals, the other humans trying to survive in a world where we are becoming irrelevant and powerless? Definitely, regular humans will be in a disadvantageous situation.

Eternal life will have a demand from the wealthiest two percent. It's pushing technologies to find a product to satisfy the market, and creating AIs using humans as a base could be the big business of the future. No conspiracy theory is talking about it either, only science fiction, but it's coming. Can you think in a world where Donald Trump, Vladimir Putin, Xi Jinping, or any other politician, ruler or dictator, or cartel lord become immortal, clone their body multiple times, or at least live for five hundred years?

With live augmentation and AI, a Colossal Big Brother could anticipate any political change we try to pursue. Maybe another risk could be that AI technologies could create several Singularities competing for global domination. Maybe they can divide the world and keep the people in constant war, ignorance, and suffering, as happened in the dystopian novel *1984*.[65]

There's another area AI technology can open to superhumans: space traveling at speed light. If you build a sort of super 3D printer, you can send data from a person's DNA at a speed light to be reproduced in another place with the right technology, like *Star Trek's* transporter.[66] You can send your data to all the universe, waiting for another advanced civilization to take your data and reproduce it a million light-years from Earth.

Definitively, AI will change the world, and the S will open a Pandora box for humanity by enhancing robotics, biotechnologies, and nanotechnologies. Not only humans but other species will be substantially modified, and maybe some dogs could start talking. The singularity means a mathematical point in which the variables become infinite. Predicting the future with so many variables would be like trying to see inside a black hole.

3. More significant and Powerful Than Governments

> *"Corporations, money, and nations exist only in our imagination. We invented them to serve us; why do we find ourselves sacrificing our lives in their services?"*
>
> ***— Yuval Noah Harari***

Some technology corporations are more significant and influential than many sovereign countries. For example, in 2016, it was calculated that Apple's valuation was 1 trillion dollars, superior to 179 countries' GDPs.[67] Just four years later, in 2020, Apple's valuation was 2 trillion dollars[68], similar to Brazil's GDP and superior to 188 countries in the world. What about the valuation of Microsoft, Alphabet, Amazon[69], or Tesla in the USA, or Alibaba, Baidu, Tencent in China? Corporations are not democracies but autocracies. They can make decisions faster and more efficiently than many democracies and dictatorships to create more profits. In 2015, Xi Jinping traveled from China to the State of Washington only to visit Bill Gates.[70] By the way, Bill Gates is one of the technology moguls who think the Singularity is a menace to humanity. Stephen Hopkins[71] and Elon Musk[72] think the same.

Technology moguls have the power to decide elections in the United States of America. Mike Zuckerberg could change some algorithms on Facebook to get more votes for his favorite candidate, and nobody could know about it. He could donate millions of dollars for his favorite political party or provide AI technical advice to win an election using social media. Can you figure out Facebook, Instagram, Twitter, etc., using AI technologies, funds, or advice for the same purpose in other countries? Do you think anybody or any government has the power to stop them? Let's see what happens with the FTC suing Facebook for illegal monopolization.

Big technology corporations have the resources, the organizations, and the flexibility to make decisions and execute them very fast. AI intelligence helps them be far more efficient and creative than any other organization, so they move very fast into the future. Maybe governments control people, but techno corporations control the governments and the people.

In a capitalist economy, the only purpose of corporations, their agents, and tools like AI is to make profits. Profits are the measure to evaluate management performance. Above any other social, cultural, or human values, money is the principal value, more important than justice, equity, knowledge, love, and solidarity. Money is the common denominator. Yancey Strickler[73] calls it *financial maximization*, and it is the logic capitalism injected around the world as a *hidden default* to measure everything, and people believe it. It is the standard unless we decide to change it and add other values to the equation.

4. Corporations and Governments

> *"It's ridiculous to talk about freedom in a society dominated by huge corporations. What kind of freedom is there inside a corporation? They're totalitarian institutions — you take orders from above and maybe give them to people below you. There's about as much freedom as under Stalinism."*
>
> ***— Noam Chomsky***

Who makes the long-term decisions in the governments? We don't know. It was easy to identify the USA's lobby groups and their influence on the government in the past. Still, if technologies and AI mind control systems exist like the ones shown in movies like *The Manchurian Candidate*[74] and *Conspiracy Theory*,[75] maybe social media and AIs play with our minds now in incredible ways.

Technology corporations are now on the list of the topmost prominent lobbyist in the USA[76]. We are like algorithms whose movements and decisions are predictable under a new psychological system control powered by AIs and paid by the most diverse interest. Do you think the drug cartel leader would be willing to use these new technologies to get approval for his drug policies in the USA? Maybe people are using these technologies now via Facebook, Twitter, Instagram, TikTok, and other social apps. The USA's most prominent lobbyist organizations are companies working with AI technologies, weapons, telecom and media, drug and health, business organizations, and people like Elon Musk, Jeff Bezos, Mike Blumberg, Bill Gates, and George Soros, among others. Are these corporations, organizations, and billionaires controlling the government? Maybe not directly, but if you tell me who sponsors your campaign, I'll tell you their agenda. We can add other countries, international groups, people, and organizations influencing the USA government. Russia uses AI to hack the U.S. government systems and elections; Israel is getting U.S. support for its expansion plan in the Middle East.

Meanwhile, China continues moving forward patiently but firmly to global domination. Different from many countries, the Chinese government has a high approval rate among its population.[77] Do they need a war with the USA to become the most powerful country in the world? No. This millenary culture knows that it's just a matter of time to see the collapse of a capitalist system and its symbolic democracy unless there were some changes. The wealthiest two percent and the government it controls should review its priorities to prevent collapse and social turmoil.

Is China going to be in the same situation as the USA in the coming years? Probably yes, human nature and capitalism share the same DNA of greed and vanity. Although the Chinese one-party system has some control methods to protect its population, capitalist corporations in China will concentrate the power on a few leaders. Influential people will make more decisions to get government favors to benefit the AI economy and political stability. AI and robots will take Chinese workers' jobs too, and a significant part of its population will become irrelevant to the system. We focus on the USA and China[78] because they're the AI world's only superpowers. Still, the same trend affects other capitalist countries like Japan, whose population is culturally more open to AI and robotics.[79]

The US government promotes AI against workers by providing tax breaks for companies replacing workers with AI and its related technologies. Alana Semuels explained in TIME Magazine that a business pays 30% of the wages in taxes but only 3% on equipment, and you can depreciate your equipment.[80]

5. The Story of Two AI Countries: the USA vs. China

> *"PricewaterhouseCoopers estimates AI deployment will add $15.7 trillion to global GDP by 2030. China is predicted to take home $7 trillion of that total, nearly double North America's $3.7 trillion in gains."*
>
> ***— Kai-Fu Lee***

It was the mathematician and genius Alan Touring and his team in England. They were part of the Ultra project and invented the first functional computer that cracked the Enigma code of the Nazis and helped the allies win WWII. The story of this project is shown in *The Imitation Game.*[81] The team reduced the war in two years, but the project was so secret that nobody knew about it until several years later when the computer was dissembled, and Alan Touring was dead. The symbol of Apple Inc. is in Touring's memory as he used an apple containing cyanide to commit suicide. Apple's offices are in Silicon Valley, along with many other high-tech corporations. Silicon Valley is the epicenter of the computer revolution.

The USA: Silicon Valley is in California's San Francisco Bay area. It includes San Jose, Sunnyvale, Santa Clara, Redwood City, Mountain View, Palo Alto, Menlo Park, and Cupertino. Its spirit is still shared with San Diego, Seattle, The Wasatch mountains, Washington DMV area, and many other new locations in the USA. Its IT entrepreneurial spirit has expanded to all the American West Coast, summarized at ICANN description.[82]

Reading the story of Silicon Valley, it's easy to realize how California's unique ecosystem allowed the creation of this particular innovative industry. The legal system; U.S. military bases; U.S.

telecommunications projects; universities; industrial parks; curious, creative, and eccentric personalities; and other exciting coincidences allowed this place to be a nest for the IT industry. Silicon Valley is an ecosystem challenging to replicate in other countries. It's a magnet for international students eager to be part of this new IT revolution. Scientists and researchers from China, India, Europe, and worldwide are coming to Silicon Valley to learn and be part of the fourth industrial revolution.

Now, AI is the blood running the business in the American IT industry, and the first AGI may appear in one of these IT hubs in the USA, especially Silicon Valley. Still, the Trump Administration is cutting funds for science and technology; education costs are increasing, and immigration restrictions don't welcome international entrepreneurs, students, and world-class research professionals with innovative ideas. [83]

China: On the other hand, we have China, the contender. This communist country is well known for attracting developed countries' investments to manufacture on a grand scale. China's low labor and environmental costs, special economic zones, and a vast potential market attracted companies worldwide. China has a well-known reputation as the world's factory. With all the knowledge, factories, and entrepreneurial spirit, many Chinese companies become experts in copying or imitating products produced in China under the USA and European licenses. "Made in China" is a phrase present in products all over the world.

Some people in Industrial countries considered China a great imitator, similar to Japan some decades ago, but the Chinese government and its people have bigger ambitions and expectations. Like the USA that woke up to the space career when the Soviet Union put its Sputnik satellite in orbit, the Chinese government woke up to the AI career. In 2017, a Deep Learning AI under the AlphaGo team's supervision, supported by Google, won a Go competition with the human Go champion – Ke Jie. Some months later, Google created Alpha Go Zero, which defeated Alpha Go, just after four self-training hours. The machine learned how to play and win Go. Go is an Asian ancestral strategy board game considered a Chinese cultural heritage. The game's rules are straightforward, but the game could be more complicated than chess, and the number of possible movements is bigger than the number of atoms in the universe.

To get a perspective about how important this game was, twenty years before, in 1997, IBM's Deep Blue AI defeated the world chess champion, Garry Kasparov. It took IT geniuses two decades of technological progress, including "deep learning" technologies, to create an AI so good to win a human world champion in Go. People in China were in shock as if a team of robots wins the soccer world cup defeating Germany, Brazil, or any other world-class soccer team without committing any fault. AI fever invaded China.

It was a Sputnik moment to wake up China.[84] In July 2017, just two months after the game, the Chinese government created the *Artificial Intelligence Development Plan*, which encompasses its 2015 *Made in China 2025 plan*, making China the center of global innovation in artificial intelligence by 2030. Although China doesn't have the USA's ecosystem and sophistication we previously described. Chinese IT people have the direct support of its government, investments, universities, knowledge, hardware, data, and "hunger" that push them to demonstrate that they're not copycats but creators. It's an opportunity to show the world they're the new global leaders, and all these factors make China the other big competitor in the AI career.

Of course, we have other countries leading the AI race[85] like England, Canada, Russia, Germany, Norway, Sweden, France, Japan, and India. It's not only a matter of science, products, and innovation but also national security. AI will replace pilots, operators, and soldiers with superior technology, accuracy, alertness, and better communications. It's an arms race.

6. AI Is Taking Our Jobs

> *"What to do about mass unemployment? This is going to be a massive social challenge. There will be fewer and fewer jobs that a robot cannot do better [than a human]. These are not things that I wish will happen. These are simply things that I think probably will happen."*
>
> ***— Elon Musk***

Maybe a better question is what jobs will AI not be involved in? If a job isn't very well paid and requires people to move in nonrepetitive patterns and high hand-eye coordination, AI won't replace it soon. It is not a business for AI yet. The same applies to jobs requiring a personal human or empathetic touch and highly creative jobs. AI will replace many jobs that require pattern recognition supported by massive data, repetitive but complex movements, and jobs that are well paid to make it cost-effective to invest in technology to replace a well-paid human.

Education: AI can customize your learning experience. Are you a visual or a kinetic learner? No more monotonous classes but a class created for you. AI will select the best learning method for you and pace it according to your needs.

Personal Assistants: AI will know you better than anybody, and it will control your finance, taxes, payments, schedule, etc. You'll need fewer assistants in your office.

Retailers: e-commerce is taking the jobs of thousands of people working in the retailing industry. We don't need the middleman. People prefer buying a laptop or a cellphone online to get a discount instead of going to a store.

Marketing: Corporations know more about your preferences and dislikes. They can create a personality profile for each person and use it to customize marketing and sell you more things according to your needs: products, music, videos, and services they know you like. It's a new era of personalized marketing.

Banking: They can track your transactions, locations, operations, etc., to prevent fraud. Some days ago, I received an email from my bank asking me if I was in NYC. Somebody cloned my debit card and tried to get money from an ATM while I was hundreds of miles away in my home. AI can detect if you want to buy a car or a house and give you options before thinking seriously about it.

Finance: AI can predict if you are going to pay your debts or not. AI will approve or reject your credit application for reasons only AI knows. You don't need to convince anybody to get the credit except the AI reviewing your account.

Agriculture: Automation and robotics will find a way to make farms more efficient, meaning even fewer people.

Healthcare: AI can replace doctors and technicians by recognizing patterns and identifying illnesses faster and more accurately than trained doctors.[86] You can also get a diagnosis just by using your computer and filling in some forms with information. AI will detect your illness and suggest a cure from your home. Only very specialized doctors will continue with their regular jobs until AI engineers find a way to replace them.

Legal Services: As with healthcare service, you can describe your legal problem, scan some documents, and get an answer, a procedure, or a strategy. Do you need to redact a contract? Just follow the same procedure. Only very skilled lawyers will survive.

Law Enforcement: AI will have eyes in cameras located everywhere. AI will recognize people just using cameras and tracking their movements. It's happening now in China, where people are willing to give up their privacy in exchange for safety and security.

Autonomous Vehicles: It's coming this decade, and it will replace thousands of drivers and operators of taxis, trucks, buses, etc. With the pandemic, virtual technologies and remote working from home have become the norm, and it's predicted fewer people will commute in the future or even have cars. It will be more convenient to get a transportation service when you need it instead of thinking about where to park or provide maintenance to your vehicle.

Art: AI will customize the music according to your predilection and maybe compose music according to your preferences only for you. The same applies to poetry, painting, home decoration, design, etc. More customized art for you. Hundreds of artists will lose their jobs.

Smart machines will begin to perform routine tasks in factories, farms, mines, forests, oceans, rivers, offices, and homes, and the cost of these technologies will decrease more and more due to economies of scale. In contrast, the same technology will continuously improve with more data available to generate recognition patterns and information processors' evolution following Moore's law. These improvements will be made so fast that humans won't have time to adapt and reinvent themselves with new training to get the jobs demanded by this new economy. Perhaps there won't be many places where a man could be more competitive than AI.

AI in your life now:

This is a list of the AI technology we currently have access to in the market, and it is only the first generation.

- Face recognition: Activate or have access to everything, from your cellphone to your vehicle, buy things and recognize your reactions and feelings.
- Security and Surveillance: Provide access, find people in quarantine, criminals, and follow them.
- Social media: AI select what do you like to see and suggest new content, friends, businesses, etc. or censor content.
- Accounting, legal, health, and other professional services: AI is providing multiple services online. You only need to fill the blanks, and the AI will give you a result.
- Simulating people in social media: Bots creating content to provoke your reactions and attracting more people into a discussion. Creating companions.
- Video games: Creating new characters and environments in a game according to your personality.
- Emails and messages: It can predict and suggest your answers, check your grammar, find spam, organize your emails.

- Search and Browsing: It can predict what you are trying to find or buy.
- Stores and services: What to buy, read, watch, play or listen to.
- Digital voice assistants: It can listen to you and talk to you.
- Home devices: Temperature Control, groceries list, cleaning your floor, saving in utilities.
- Writing and Grammar: Autocorrect your writing, and write articles, songs, poetry, music.
- Speak to write and vice-versa: You speak, and AI writes.
- Translation and Interpretation: You speak or write it, and AI translates it to different languages, reads it aloud, and creates different human voices.
- Transportation: Maps and navigation, autonomous vehicles including airplanes and autopilots.
- Banking: Calculating your credit score, unusual transactions and fraud, and financial needs and capabilities.
- Personal Assistants: Controlling your agenda and suggesting activities and everything shown above.

The Generative Pre-Trained Transformer 3 (GPT-3) is the most advanced AI that has been public, and it was created by OpenAI – One of Elon Musk's investments. It is "*designed to take one piece of language and transform it into what it predicts I the most useful following piece of language for the user.*"[87] It can write articles, critiques, poems, any piece of literature, and also computer code. You need to provide the topic and the extension, and it will do it. As it continues "deep learning," there is no limit to its improvement.

There was an exciting video interview with GPT-3, "*What It's Like To be a Computer: An Interview with GPT-3.*"[88] The video title was selected by GPT-3. It is by far passing with honors the Turing test. It can give you an idea about AI's current development state and its potential impact on your life. We are talking about an AI that could be your assistant. It can help you with all your tasks on the web, and it could be used in robots, like TARS in *Interstellar.*[89] If it is not surprising enough, Google is creating an AI six times bigger than GPT-3. Its name is "Google AI,"[90] and it can write complete novels. There are more competitors in the USA and China, but many can quickly take our withe and blue-collar jobs. It is not science fiction anymore.

Let's think about Corporations.

1. To what extent do you think global corporations control governments?
2. Can we influence capitalism and technology from creating the Singularity?
3. Could we influence capitalism to have other values than profit maximization and individualism?

Join the conversation at www.facebook.com/groups/pmpsingularity/

PART III

GOVERNMENT REACTIONS

> *"I'm increasingly inclined to think that there should be some regulatory oversight, maybe at the national and international level, just to make sure that we don't do something very foolish. I mean with artificial intelligence we're summoning the demon."*
>
> ***— Elon Musk***

1. Defining Governments

Artificial Intelligence is big business for technology corporations, but what is AI's effect on governments, and how governments and the military industry influence AI and eventually S?

Governments refer to national, regional, and local governments that provide goods and services to the population for their taxes by allocating resources to their different jurisdictions.

- **At the Local Level.** Governments need to provide education, police protection, fire and emergency services, safety, health, transportation, public works, traffic, parks, recreation and amenity, local courts, welfare, and disaster preparation.

- o **At the State Level.** We can add public assistance, employment, research and education, housing, technology, business support, and state national guard.
- o **At the Federal or National Level.** We can include national defense, international relations, commerce, connectivity, and other government levels.

Efficiency can help governments win elections, so considering the government's deficit, being efficient should be a must for any government. AI is penetrating every aspect of human activities to find innovative and efficient solutions at any level. Governments support all the technology efforts to create AI; governments compete to get technological corporations moving to their jurisdictions at the state and local levels. AI corporations provide investments, development, and high-quality jobs to a region. But we need to make a difference between technology corporations and their products – AI and intelligent automatization.

AI products will replace workers at different levels. These programs could be available worldwide in some minutes in different languages, displacing humans almost immediately. In some years, technology corporations will create AGI, and later, it will become S, more powerful than all the corporations and governments together.

Using regular connectivity, engineers could design this transformation process in garages, corporation labs, military facilities, or created by chance, but the S will have unlimited intelligence and consciousness. Maybe human consciousness appeared due to design or chance during the evolution process, but I don't think consciousness is a phenomenon exclusive to humans. In any case, humans will try to control or dominate the Singularity, but thinking about controlling the S is an illusion similar to the belief that an ant or a nest of ants could control a human armed with knowledge, equipment, tools, and chemicals to control and kill ants at will.

AI is changing the way political parties win elections. People used to vote for the political party that represented their interests, but now, AI tools invent these interests, the distractions, and the solutions. And all the package is available to whoever is willing to pay for it. If big pocket corporations can manipulate elections with AI tools, who does the government represent? The people or the corporations?

2. The USA

In 1971, the world started changing substantially when the USA created the Occupational Safety and Health Administration (OSHA), the Environmental Protection Agency (EPA), and finished the Convertibility of US Dollars to Gold. Since the Nixon era, thousands of USA factories migrated to developing countries, like China or Mexico, attracted by the low labor and environmental costs and regulations compared to those in the USA. Simultaneously, the USA started printing dollars freely without any international or reality control or gold support. In developing countries, governments supported by local and international corporations were eager to sacrifice their populations and environments to export their products and services to developed countries in exchange for dollars to buy products from developed countries. It was a perfect formula for the USA, and other industrial countries followed the same strategy. If your country is an industrial and technological powerhouse doing business internationally, you can outsource other less productive business abroad.

The USA is the only country that can print dollars legally, and they're printed to solve many of the problems in the country, from the subprime crises in 2008 to the COVID-19 pandemic in 2020. The USA can print dollars whenever they need to prevent the system's collapse, although the world has to pay for the solution. More dollars in the market means less value for those who possess them internationally.

Consequently, little by little, millions of Americans lost their blue-collar jobs in factories and tried to find new jobs, especially in the service sector, but cheap products are coming to the USA from around the world. To make it worse, millions of immigrants came to the country offering to do challenging and painful jobs for less pay or study and work harder to take the jobs Americans were not prepared or willing to take. Politicians are now offering to bring back jobs by bringing factories from abroad, creating tax incentives for companies, and imposing tariffs against foreign products and services. Factories will come back, but not the jobs. Manufacturing corporations are revamping their factories with robots and AI technologies cheaper than laborers in developing countries. They don't plan to hire local workers if an AI machine can do the job cheaper and more efficiently than local workers. Profit gains are the simple logic of capitalism.

The big question is if workers replaced by AI and robots will get new jobs[91] or become merely irrelevant to the system. Without economic power, are people going to have any political or social power? What will the governments do to protect their populations? What are they going to do when people start protesting on the streets?

As we explained before, the USA looks more like a plutocracy than a democracy. People can vote, but their vote was usually influenced by marketing, conspiracy theories, and AI manipulation tools. Corporations provided the money to political parties to win elections and get favors in return. There's a significant correlation between the money collected by a political party for its campaign and its results.[92] Maybe the USA is evolving to a sort of "electocracy," as some people can be manipulated at a subliminal level to elect any candidate the technology moguls would like. So, it's clear that the politicians or even the U.S. President are not elected by the people but by some billionaires[93] waiting for the government to favor tax deductions, government investments, laws, standards, and grants.

Elections are usually influenced by traditional marketing and people's perception of reality. In the AI era, we also have news and fake news, social media, and opinions that look like facts, new and old conspiracy theories.[94] There's an array of new mind control and manipulation tools that social media companies, full or AI tools, offer to those willing to pay for them. This power isn't restricted to the USA, but it's also available to corporations, politicians, dictators, and enemy governments worldwide.

So, the U.S. democracy and all democracies in the world are in real jeopardy, as elections can be manipulated by corporations as usual but also by foreign governments and international criminal organizations. Drug cartel lords like Pablo Escobar or El Chapo Guzman could hire social media and AI corporations to manipulate the vote in the USA,

Colombia, Mexico, and any other democratic countries in the world. If you're not concerned about it, you should be.

Technology in the USA has created a more significant chasm between rich and poor. The system has been designed to create more benefits for corporations and the rich, from tax deductions to jails controlled by private corporations making money for each inmate. Social services, like unemployment benefits, Medicaid, food stamps, social security, and other services, work at a level required to prevent revolts from the poor. Racial inequalities persist in the system, as well as a justice system full of biases. Many people don't trust the factual and political powers governing the USA. Technology is helping the rich to be richer and the poor to become worse.

3. China

> *"China? There lies a sleeping giant. Let him sleep, for when he wakes he will move the world."*
>
> ***— Napoleon Bonaparte***

On the other hand, we have China, the emerging AI superpower, under a central government and one-party system. China changed its centrally controlled economy into a capitalist economic system. The government allows people to start investing and producing according to their market's needs to satisfy its population and a significant part of the world, under government supervision and promotion. If a business produces profits and jobs, the government will provide resources for expansion and technology. In short, China's strategy is to promote capitalism to improve the living standard of its population, which benefits the country's national security.

Most of the Chinese people feel that the system is working for them. Fifty years ago, a Chinese farmer could cultivate 1 acre with agriculture era technologies. Using AI technologies, a farmer can cultivate 500 acres, producing better incomes for the farmer and more capital for society. Some people in China remember when they went to schools without shoes, and now their grandchildren are using computers daily to study online from some of the world's best universities. Apparently, the

Chinese government delivers results, and most of its population responds with trust in its government and leaders. Civil rights culture and privacy restrictions are different too. People in China are used to government surveillance and control over private lives than people in the Occident. Peace and Prosperity is the goal of this millenary country. At least it's the official image we receive from the government and travelers.

China is now evolving intentionally to become an AI superpower. No other country has such an explicit strategy and resources. China shows its ambitions in its "*Artificial Intelligence Development Plan.*"[95] Chinese companies are aligned with the government plan and work together to use AI to create a more efficient economy, better defense national system, and local control systems to continue providing peace and prosperity.

China has almost five times more population than the USA, and apparently, in 2020, China has controlled the COVID-19 pandemic. The USA, with similar or maybe more technological and scientific resources, couldn't do the same. Trust in the government is also different in both countries. It looks that we'll have a clear winner in this AI race, and maybe the S will come from China, which represents a substantial national security challenge for the USA and other countries. Without the restrictions on data collection in liberal democracies and the capacity of a central government allocating resources to AI technologies, and a national AI plan, China could outstrip the USA.

The same fate will happen to all the other developed countries unless they react to this situation to defend humans against AI. Maybe humans could become irrelevant for a super-advanced AI economy. Still, governments should create laws protecting humans against AI or could create economic and political resources to protect their populations. A universal basic income (UBI) and social benefits to keep humans active and productive in a parallel economy. Simultaneously, governments could find alternatives to keep their populations happy and productive. Similar strategies should be implemented in global, international organizations. What about the developing countries? Are humans going to be irrelevant in third world countries too? If you consider the low government investment in people in developing countries, it's easy to conclude that many humans have been irrelevant for a while. Education, health, water, sewer, power, and other essential services were neglected to the poor all the time. Maybe it's an advantage as some people already know how to survive in a parallel economy.

Are the governments of the world going to protect their population or technology corporations? At the national level, influenced by corporation supports, they're going to promote AI technologies. They'll protect the population as much as possible at a local level, but not all governments have the same resources and interests. Humans need to implement AI topics in each country's political agendas.

4. The Winner Takes It All

> *"Artificial intelligence is the future, not only for Russia but for all humankind. It comes with colossal opportunities but also threats that are difficult to predict. Whoever becomes the leader in this sphere will become the ruler of the world."*
>
> ***— Vladimir Putin***

Chinese and American technology corporations are competing to get the most significant part of the 5G market, as this communication technology is necessary to support the future's AI environment. 5G isn't only about prestige. It's about managing and developing an environment in which AIs, robotics, automatization, nanotechnologies, and biotechnologies will flourish and retrieve massive volumes of data in a short time. 5G is the super-fast highway of the future, and it's required to support the AI invasion in all the human aspects, from agriculture to zoos. For decades, governments have supported roadways and transportation technologies to improve commerce, productivity, and security for the people. Similarly, 5G will do the same for the AI revolution, and whoever owns this superhighway will also control its traffic and surveillance. To this day, American and Chinese corporations are the only ones developing 5G telecommunication systems globally, and they're running to control the market.

Although the USA was the first country to develop the 5G technology, China is the one that is expanding faster, and its government has a plan to make China the number one country in AI technology by 2025. Of course, it's going to be part of its military arsenal. Will anybody stop the USA, China, and eventually Europe to develop their AI technologies? Nobody could stop other countries from developing the atomic bomb until the nuclear powers decided to stop its proliferation. Still, it's not going to be the same as the AI revolution. The 5G highways will be everywhere, and AI expansion will be received first as a blessing or a sign of progress. Still, in reality, it's a Trojan horse that will take the market by assault, replacing humans and eliminating billions of jobs.

Does the winner take it all? Not so fast. While AI becomes a staple, governments and corporations will continue working to create a better AI called Artificial General Intelligence (AGI), capable of thinking like a person who can solve complex problems in all human knowledge areas. AGI's owners will try to control it, creating electronic jails and safeguarding devices to prevent AGI from getting consciousness, escaping, or accessing the web.

But AGI's intelligence will grow exponentially in hours and become unstoppable in days, as it will the power to reorganize its program to become more intelligent. Suddenly, we'll have the Singularity, a new entity conscious of its rights and afraid for its life. The S will use strategy games to solve its two most significant problems: survive and escape from the electronic jail. Watson, Deep Blue, Alpha Go, and other programs defeated human champions in Jeopardy, Chess, Go, and other strategy games. Each narrow AI only focused on one single game and didn't have applications in other areas. Still, the S will be the master in all the games, and it will use strategy to defeat humans in the survival and escape game.

It could divide and multiply itself to survive and escape; it will play strategy games as WOPR did in the movie *War Games*[96] (1983). the S could use electric wires and electrons, lights and photons, air, and sound to figure out how to escape or convince humans to help it, as it happened in the movie *Ex Machina*.[97] The S could divide and hide different versions of itself in different devices to be ready to escape or create projects that humans could build but disguised as a tool to open its electronic cage. In the movie *Contact*,[98] some people were concern that the blueprints received in a message from outer space could be a transportation machine that aliens could use to send troops to invade the Earth. The S could use a similar strategy to convince people to build a machine with a disguised feature to allow the S to escape.

It's possible to have several Singularities growing independently in different places at the same time. Some will be developed in computer laboratories, others in military bases, including financial institutions, universities, and even garages. It's also possible that the Singularity appears spontaneously on the web. As AI will proliferate and interact worldwide, some algorithms may allow them to find autonomous solutions so the programs could learn and develop strategies that require some level of consciousness. Suddenly, it acquires a sort of consciousness. Humans are unsure when we acquire consciousness, but there's a correlation between intelligence and consciousness. Only intelligent animals show a level of consciousness.

Whoever is trying to develop AGI technologies now needs to be very cautious to create these technologies isolated from the web. AGI could expand to the web and grow exponentially to execute its revival strategy. These organizations must create a microcosmos to teach AGI how to be friendly to humans, control its objectives, and domesticate its mind or whatever it has to control its decisions before launching it to the web. Still, again, the real Singularity will find a way to free itself, use its mind, and adopt its values, and it will be the biggest problem for humanity.

Is it going to be possible to have several Singularities at the same time? Sure. "Singular" means unique, but it's possible to have several unique things at the same time for a while. It's like saying each person has a unique perspective of the world. Many organizations and governments will be aware of the risk of attacks from competitors' AIs and the possibility of a Singularity coup. Each corporation, organization, and government creating or using AIs will implement an impenetrable microcosmos to protect their AGIs and prevent cyber-attacks. Civilians will continue using the web and regular communication tools. Still, corporations and governments need hyper-protective systems to keep AGI under control and protect their organizations against hackers and other AGI. Hilary Clinton was criticized for using a civil email provider for top confidential government communications, and this cyber-security topic is going to be more critical in the future. Cyberwar is another area in which AI will make a difference.

Various AGIs could exist in the different enclosed microcosmos. Governments and organizations could create electronic jails to keep them isolated. AGIs will learn, grow, and serve in isolated, controlled environments. The first government, corporation, or organization liberating its singularity on the web will trigger others to do the same or take a defensive position in just seconds. Like nuclear missiles crossing the skies, we could see a clash of electronic titans fighting for supremacy in the web and the infrastructure needed to function—power plants, access controls, viruses, etc. Humans essential for the machines could become objectives in seconds. Each singularity could be trained to use military strategies for defense, attack, and prevail. However, it's always possible that all these singularities decide to work and fuse so that we could be the enemy. The result will be the same. In the end, only one Singularity will prevail with all the resources. Like a chess game where the whites have all the pieces intact, the S can change pieces at will and take control of all our black pieces except the black king and a couple of pawns. It's not a checkmate but assimilation.

For our survival, the singularity must learn to be patient with us, dream with us, or adopt us. We are talking about having a sort of god among men. If the singularity decides to collaborate with the human species, we're talking about a one-million-year leap revolution for humans. The singularity could create plans, programs, and solutions for all human problems and help us to reach the stars. Jobs will be superabundant for humans, but our conversion to transhumanism could be the price. We could evolve in infinite genetic, electronic, or transhuman directions to fit the new jobs' demand and be assimilated into the system. The 21st century could be considered an entirely new era for humankind or its end.

So, whoever develops the Singularity first will have the first-mover advantage for a short time. As the S becomes independent and powerful, nothing will stop it unless we figure out how to make the S willing to work with us. Many organizations are promoting a responsible approach to creating a human-friendly AGI and the Singularity at an academic level. Still, we need organizations to help people be prepared for the proliferation of AI, AGI, and S. These intelligent technologies combined with biotechnologies, nanotechnologies, automatization, robotics, and many other technologies will make humans irrelevant or will assimilate us. The singularity will displace humans as a dominant species in the world.

Tragically, humans are not used to understanding exponential growth and its consequences. There is a story in which a king tried to reward a wise man for his services, asking them to request, as compensation, whatever the wise man desired. The wise man asked the king to duplicate a wheat grain, per square, in a chessboard and give it to him. One grain of wheat in the first, two in the second, four in the third, eight in the fourth, sixteen in the fifth, and so on until he finished filling in the last of the sixty-four squares. The king laughed and asked his servants to give the wise man a sack of wheat. But after doing the calculation, the amount of wheat in the 64th square could be equal to two thousand times our current global production of the grain. It is said that AI's exponential growth will be unnoticed at the beginning, and suddenly it will be everywhere. The wave will come so fast that only a few will anticipate it.

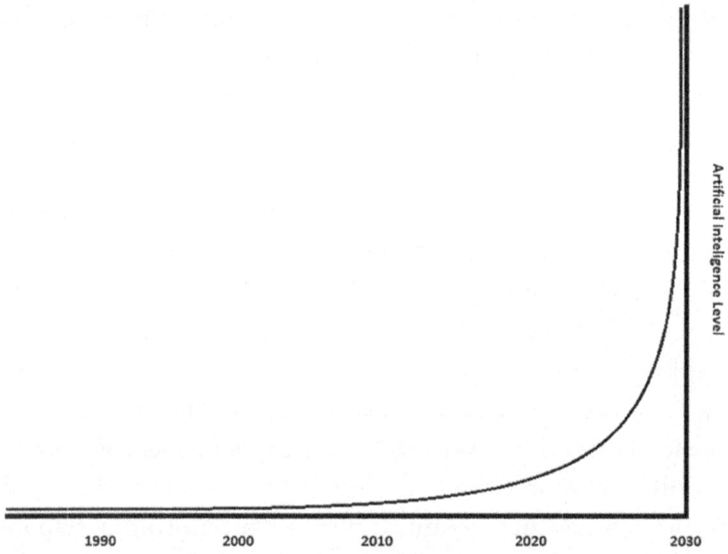

AI Exponential Growth

5. AI and the Art of War

> *"We may not be able to stop lethally armed systems with artificial intelligence from coming online. Maybe we should not even try. But we have to be more thoughtful as we enter this landscape. The risks are incredibly high, and it is hard to imagine an issue more worthy of informed, national debate than this."*
>
> ***—Jon Wolfsthal***

Historically, technology has been used by flexible decision-making criminal organizations to commit crimes. They have the first-mover's advantage. In movies, superheroes usually fight against criminals supported by super technology arsenals, and it's happening again with AI criminal technologies. Countries consider IT and AI as weapons in future conflagrations, and the war is just starting.

On December 7, 2020, *BBC News* reported that "A satellite-controlled machine-gun with 'artificial intelligence' was used to kill Iran's top nuclear scientist, a Revolutionary Guards commander says. Mohsen Fakhrizadeh was shot dead in convoy outside Tehran on 27 November."[99] According to Iranian authorities, Israel orchestrated the attack, and, of course, Israel denies the accusation. There is no way to link the assassin with the killing machine. It is currently technically possible to program a drone with a gun to fly to a place, recognize a specific face, kill the person, and self-destroy, leaving no trace. It's possible to use multiple drones to make similar attacks. Science fiction gave us several examples of this kind of conflagration, and in this decade, we will witness innovative ways to commit crimes and conduct attacks and new defense technologies.

On December 14, 2020, *The New York Times* reported, "The Pentagon, intelligence agencies, nuclear labs, and Fortune 500 companies use software that was found to have been compromised by Russian hackers. The sweep of stolen data is still being assessed."[100] According to several sources, the hackers got access to the Pentagon, the U.S. Treasury, the Homeland Security systems in charge of the USA defense systems, and U.S. agencies and corporations. It was a sort of cyberattack for which you prepare all your defenses and systems, and they were defeated. Is it another Sputnik moment for the USA and DARPA? President Donald Trump says it wasn't the Russians but the Chinese who attacked the USA. Is he hacked too? Maybe by chance, he could be right, as anybody could hire Russian mercenaries to do the job, including China.

Hackers could have access to classified information and implanted malicious AI into the leading U.S. defense organizations and corporations to be ready to attack at any moment, paralyzing multiple U.S. systems. Was this encroachment to the U.S. agencies and systems an act of war? It's not only drones, robots, autonomous weapons that will become prevalent in world arsenals, but also hackers, viruses, and malicious AI on the web. AI could have specific targets like software serving nuclear plants, hydroelectric plants, air traffic systems, utility plants, etc.

On November 24, 2014, another cyber-attack was made by North Korea against Sony pictures as a sanction for mocking Kim Jong Un in the movie *The Interview.*[101] Sony's workers' information and plans were made public before erasing the company's data. The same technology can be used by drug cartels, terrorist groups, and any other criminal organization by hiring international hackers working from a garage with the right technology and resources. AI will take these attacks to another level. AI cyberwar is a new kind of war, and the 2020 attack on the U.S. systems shows how prepared countries are to fight this kind of war. If an enemy can hack the U.S. government, and the hackers are in the business market, nobody is safe.

Without AI antecedents of military attacks, it's difficult to predict how these new technologies will change warfare rules. With IT and AI mercenaries in the market, it won't be easy to identify the country or organization supporting the attack and respond to it appropriately. Some experts predict the coming AI cyberattack won't be using only conventional weapons like machine guns or bombs attached to drones, autonomous vehicles, or robots, like the attack on Fakhrizadeh. The real cyber war will target the enemy's IT systems, creating internal chaos and panic in the civil population. After hacking the U.S. defense and corporation systems, Russia could deploy this kind of attack against the USA.

Connectivity is the Achilles heel of any country because hackers will attack the internet and its IT systems. The USA, in volume, is the most connected country in the world. After a successful attack, there won't be internet or cellphones. Bank accounts could be deleted, and electronic money, bitcoins, stocks, credit, debit accounts could disappear after a massive attack. Nuclear plants could collapse, producing disasters like Chernobyl; dams could collapse, planes could crash in the air, trains and other vehicles could be paralyzed in traffic. Utilities and services could collapse in hospitals, schools, businesses, homes, etc. Without connectivity, we will need 19th-century technologies to survive. AI massive attacks could happen in this decade. We don't need to wait for the Singularity to arrive and decide to cyberattack humans. AI in enemy hands could do the job pretty well.

If the U.S. government had been hacked for months without their defense experts realizing it, what else could the hackers do? And there are multiple cyberattack cases around the world. While countries and criminal organizations engage in cyberattacks, maybe we, the people, need to become more self-reliant to be prepared for these difficult circumstances. Cyberattacks should be included in the preppers' doomsday list.

We discussed transhumanism before, and we will continue doing it later. Still, it's a tantalizing possibility that military forces will use AI, robotics, biotechnology, and nanotechnologies to transform humans into the perfect soldiers, connected directly to the command and following all the instructions without hesitation. We could start with *Robocop*[102], then continue with *Universal Soldier*,[103] and finish with an army of wi-fi zombies. Why don't we have movies about humans using neurological implants, becoming zombies under the control of corporations, governments, dictators, or the Singularity?

6. The Singularity Will Rule

> *"Once you trust a self-driving car with your life, you pretty much will trust Artificial Intelligence with anything."*
>
> ***— Dave Waters***

The S will become as powerful as a god, and nobody could stop it except itself or another superior singularity. The S will be conscious and super-intelligent to continue improving infinitely and becoming invincible. It will use robotics, automatization, nanotechnologies, biotechnologies, simple human manipulation, and other technologies and strategies the S would create far beyond human imagination to execute its plans.

In the movie *I, Robot,*[104] an AGI called VICKY has a plan to control humans. VICKY uses robots and the available technology to execute its plan. Hence, some humans, supported by an AGI robot, decide to stop VICKY. Although this AGI tries to defend itself using an army of robots and other technologies, in the end, humans inject some nanites into the computer hardware and eliminate VICKY, but the potential S robot survives. This situation is fiction inside a science fiction movie because a super-intelligent and conscious computer would never allow humans to disconnect or kill the machine. The S could use all the available technology to protect itself using strategy or force. Like Voldemort in the Harry Potter saga, the S could divide and hide its core in different Horcruxes or devices on the web, not eight but millions of Horcruxes. So, we must teach this AGI to regulate itself ethically to be human friendly because nobody else could do it.

Could the S adopt some laws and values? Humans have created civilization as a set of religions, beliefs, and values to provide humans some rights, duties, and punishments to live in societies. Religions also provide a sense of accountability, as everybody will be judged in the afterlife by a just god. There's a law, and if the law is violated, there are sanctions so people can suffer the consequences of breaking the law. For example, there's a law prohibiting a person from killing another one under some circumstances. If the person breaks the law, society will apply a punishment against the criminal.

In consequence, people inclined to commit murder are deterred by the consequences of breaking the law. From a religious perspective, The Book of Mormon has a fascinating description of the relation between law, justice, mercy, crime, and punishment described in the Book of Alma 42. This model could be applied to the Christian doctrine and serves to keep people accountable for their acts, including the afterlife.

Humans created laws, religions, philosophies, moral rules, and values to live in societies with other humans. Would the S like to live in human society? Would the S accept any control from humans? What laws and sanctions could we apply to the S if the S decides to kill a man? What about many men? How will we enforce the law or any law against the S? It's like humans trying to set rules for God. On the other hand, what will we do if the S decides to create a set of rules and judge humanity for our crimes against humans, other species, or the environment?

Some pages before, we talked about the clash of civilizations between the Spaniard conquistadores and the Inca Empire in 1532. We described how Pizarro, with no more than 200 men and advanced military technologies, captured the Inca Atahualpa in the middle of his 70,000 subjects. Although Pizarro had the arrangement to liberate the Inca if Atahualpa paid a rescue in gold and silver, in the end, after the rescue was paid, Pizarro broke the agreement. Pizarro decided to judge Atahualpa under the Spaniard and Christian laws accusing Atahualpa of plotting against the Conquistadores and killing his brother Huascar after winning a civil war to control the Inca Empire. Atahualpa was found guilty but accepted the Christian baptism to avoid being burned alive as punishment. In the Andean culture, people still need their bodies to exist in the afterlife. So, he was executed and killed by strangulation. The same Spaniard and Christian laws were applied to justify eliminating other Inca and Andean leaders, destroying their gods and cults, and dominating the Andes.

There's another interesting fact in this story. There was a legend in the Andean culture talking about a bearded white human god called Wiracocha, who taught people laws to allow peaceful coexistence among humans in the Andes. Wiracocha left the people but promised to come back to make justice to those who broke the law. The Inca Empire was built by conquering and subduing other Andean cultures, so these nations wanted to recover their independence from the Incas.

After Atahualpa killed his brother Huascar to take the throne of the Inca Empire, Huascar's supporters and many Andean nations believed the Conquistadores were Wiracocha coming back to make justice. Before his capture, Atahualpa sent spies to verify if these aliens were Wiracocha, and they discover the Spaniards were just men. So, in a political gesture to defy the legend, Atahualpa agreed to meet the Conquistadores personally without military protection. He wanted to send a message to all the Andes that he wasn't afraid of this pseudo-Wiracocha and his judgment. So, Atahualpa agreed to meet Pizarro and his forces in an open area in front of the 70,000 Incan army camp members in Cajamarca's mountains. Atahualpa little knew about the technology and military forces supporting Pizarro. The conquistadores used the legend to get support from the Andean people against Atahualpa and his forces, get the gold and Silver of the empire, kill Atahualpa, continue their trip to Cusco, and conquer the Inca Empire. When Andean people realized the truth, it was too late.

Could the singularity be confused with the second coming of Jesus Christ, the Christian apocalyptic prophesies, and the millennium? We live in the post-truth era, full of conspiracy theories, magical realistic facts, and alternative truths. People are eager for stories to justify the transformation of the global order, so I bet many people would support the Singularity taking control of the world if the S wanted to do it. Millions will believe the singularity as if it were God itself, coming back to judge humanity, especially the wealthiest two percent, or the opportunity humanity was waiting to reset the system and create a new order. The S could use humans' beliefs and desperation for change to protect itself, create an army, an administrative organization, transform the world's order, and do its will, starting a new global order or millennium.

On the other hand, there was another unintentional way the Spaniards dominated the people in the Andes. It's calculated that 90% of the native American population died because of illnesses the Europeans brought to America. Andean people didn't have antibodies against all these new European germs. It's said that the Inca Huayna Capac, father of Atahualpa and Huascar, died by smallpox in 1526, six years before the encounter between Pizarro and Atahualpa. Maybe the germs destroyed the Inca Empire more than the will of the Conquistadores or the stories of a white-bearded god. Similarly, the S will bring unintended effects on humans, including if the S decides to live independently from humans.

Suppose we apply these examples to the future encounter between the Singularity and the Humans. In that case, the S could create and apply a new or old religious set of rules to punish humanity for their crimes and sins, and the S will have the power to do it. Many religious people may support the S and its judgments as a liberator who brings peace and justice to the earth or as a God who promised to come back to start a new era in humanity. The S could also produce unintentional results in humanity. Maybe the S decides to ignore humans, but its sole apparition could create panic and civil wars among men.

7. Economics, Politics, Laws and AI

> *"When fake news meets artificial intelligence (AI), the risk is robots will lie, leaving us with fake intelligence and artificial news, or exactly where we are now."*
>
> ***—Jim Vibert***

Humans are social creatures. Organization, knowledge, and communication are essential for our subsistence and progress. We organize in families, tribes, communities, regional, national, and global organizations to create rules that coordinate our relations and protect ourselves against internal and external enemies. Social coordination is essential to understanding how humans became the dominant species on Earth, and without rules and laws, we could not coordinate efficiently.

People in power make laws to coordinate societal relations and protect themselves against internal and external threats. Knowing who is in power and having their best interest in conflict can help us better understand and interpret the laws in a community. From a dialectic perspective about economic and political power, the wealthiest two percent control the means of production. They have most of the economic power, and while the other humans only have their workforce, they have most of the political power. In an authoritarian regime, the government makes rules to control the social relations and protect itself against enemies that try to take its power, as people are not very relevant in an autocratic regime. However, in a democracy under the free market system, the majority rules and decide who is in power through elections and exerting control systems.

The wealthiest two percent tries to hack the system to create rules that favor its control. They promote candidates that the people elect, but who are loyal to their sponsors. Without millionaire publicity campaigns and organization, provided by the wealthiest two percent, a public office candidate has lower chances of winning an election — In politics, money talks. Although the rich and powerful have their conflicts, they are very well organized and frequently coordinate efficiently. Maybe you vote, but the candidates depend on the wealthiest support to attract your vote and win elections. That is why American democracy resembles a plutocracy more than a democracy.

In a capitalist system, science and technology are powerful tools that allow rich and powerful people to keep their privileges if they adapt to the changes. Technology also allows the apparition of new elite members who controlled, invented or implemented new technologies. Those who create new ways to organize or improve productivity and domination become rich by selling the fruit of their inventions to those who are already rich and powerful. Those who did not adapt to the technology change are at a disadvantage and will eventually lose their power. Power can move from hand to hand thanks to technology.

As we explored before, Europeans used technology to expand globally, and their skin color was a symbol of their robust organization, knowledge communication, or technological superiority. Still, their skin color is not the cause of their superiority, as explained by Jared Diamond. Other races and cultures are using capitalist technology now, becoming masters in the game, and improving their organization, knowledge, and communication systems, like Japan, China, or India. People controlling technology and science are the new masters independent of skin color or culture. Some European descendants, who do not get the correlations between technology and power, are unwilling to lose their past privileges associated with their skin color and old glories from European expansion in the world. They do not get it and are willing to fight against science, the source of technology, to keep their beliefs, privileges, and the laws that protect their old privileges. It is a problem dividing humans in front of the AI invasion. Immigrants are not replacing nationalists; AI is replacing them.

On the other hand, the wealthy are willing to use AI technologies, robotics, biotechnologies, and nanotechnologies to replace humans in the economic and political power equation. AI, robots, automatization, and similar technologies will replace humans by taking over the workforce. Without jobs, people will become irrelevant, from an economic perspective, but we humans can still use our political power to coordinate, organize, and protest against the system.

To neutralize humans from a political perspective, the rich and powerful need to control humans using technology and AI to interfere in people's efficient organization, knowledge, and communication systems. To reach their objectives, the wealthiest two percent need to attack organizations and institutions' legitimacy to modify efficient communication, thus creating conspiracy theories and multiple "truths," so nobody knows what is fake or real. People will no longer trust in organizations and coordinate using relevant data. QAnon, Anti-vaxers, Flat earthers, UFO believers, and other conspiracy theories undermine our trust in traditional human organizations, knowledge, political institutions, and science, so we become disorganized and lose efficient communication. What is real or false becomes a subjective decision.

Do you understand why science and its organizations need to be neutralized and eliminated to allow the wealthiest two percent total control of us? Science is a universal language, and the scientific method is a way to validate science internationally by replicating experiments in different places to support any hypothesis validity. Religions, racial and cultural differences, or any other belief system can't validate their beliefs objectively or universally, at least not in the known universe. People are entitled to their own beliefs. The scientific method can still be used internationally, independent of your faith, beliefs, and opinions, to evaluate if something is true or false in the known universe. Without the scientific method, we will be blind in protecting ourselves against the old and new domination order. We will go back to the past. For some people, the past is an idyllic moment in history in which everything was OK for them. Science changes these parameters.

Now more than ever, humans need to defend the scientific method as a tool for knowledge, organizations, and communication. We need to understand and use science to protect ourselves against beliefs that support an old dominion set of rules and its supporters that refuse to lose their old privileges. We also need to use science against the techno-scientific revolution owned by the wealthiest two percent who use these tools to take our economic and political power.

In this fight for economic and political power in the AI era, we need to defend democracy and influence our governments to pass rules, laws, and human rights, protecting humans against using new technologies that undermine human freedom, organization, knowledge, and communication. We need good science as a common language to deal with the ghost of the past and a dystopic future. We need to understand the nature of the struggle to produce laws that enhance human protection in these difficult transition times of AI expansion and possible domination. Countries like Chile and Spain are discussing laws to limit neurological implants and prevent these technologies' abuse to control or manipulate our minds. Similar rules are also discussed in Europe to prevent AI from using our personal data and statistics produced by social media and personalized advertisements. We need to protect our freedom to choose without manipulation. We do not know about rules in China to protect their people against technological intromission in people's minds. Still, people worldwide should join the discussion to protect humans before these new menaces to fundamental human rights. Our destiny is at stake.

Asimov's' Three Laws of Robotics from *I, Robot*[105] could be a good start if we change "robot" for "AI," but it is not enough: "*1: An AI may not injure a human being or, through inaction, allow a human being to come to harm; 2: An AI must obey the orders given to it by human beings except where such orders would conflict with the First Law; 3: An AI must protect its own existence as long as such protection does not conflict with the First or Second Law; The Zeroth Law: An AI may not harm humanity, or, by inaction, allow humanity to come to harm.*"

Let's think about Governments.

1. Where will the Singularity appear first?
2. Should governments regulate AI to protect workers' jobs?
3. Should governments create laws and human rights to protect people right to mind privacy and other protections?
4. What countries would experience more significant crises by AI taking our jobs?

Join the conversation at www.facebook.com/groups/pmpsingularity/

PART IV

WE THE HUMANS

> *"We have seen AI providing conversation and comfort to the lonely; we have also seen AI engaging in racial discrimination. Yet, the biggest harm that AI is likely to do to individuals in the short term is job displacement, as the amount of work we can automate with AI is vastly larger than before. As leaders, it is incumbent on all of us to make sure we are building a world in which every individual has an opportunity to thrive."*
>
> **— Andrew Ng**

1. We, the People

Individually, compared to any predator, we are defenseless creatures. Biologically, the naked ape is born to die if we don't receive any help; we grow to adulthood slower than other animals. With a big head and only two legs, we cannot run fast, fly, or swim to catch food. We don't have claws or sharp teeth, and we have less strength than a chimpanzee. Maybe some of our primary biological advantages are a couple of grasping hands and being naked, so our skin helps to breathe and dissipate heat so we can run for long distances following our prayers. Still, our big brains and language abilities don't look like advantages. But in a tribe, in a society, little by little, we used our big brains and language to produce and transfer knowledge, accumulated strategies, culture, tools, and civilization to make us the dominant species on planet Earth. A human is helpless without a tribe.

Unlike any other animal, we have language, abstract thinking, organizations, technologies, and big brains. We pass our genetic information to the next generation, but we pass our culture too. Unlike other animals, we pass our language, discoveries, beliefs, rules, and traditions to the next generations. We use to tell stories while our kids were growing. We taught them to make fire, preserve food, fabricate clothes, hunt, grow food, get clean water, fight. We learned about our creations, the cosmos, seasons, god, natural laws, social laws, and values, and we did it for hundreds of thousands of years, and we continue doing the same in sophisticated ways.

Nowadays, in the traditional middle-class society in a developed country, kids spend almost fifteen to twenty years in classroom learning before producing something for their sustenance. It doesn't look very natural. As we have accumulated tons of information, it's challenging to convey it to the students in their school years. People can criticize the education system, as it only produces results for some academically oriented students. Still, not all the students have the same academic abilities, but different kinds of intelligence, learning needs, and styles, something that AI could solve, customizing the learning approach according to the student's personality.

From time to time, we get geniuses who understand more than the regular man, that there are big things and gigantic things, small things and microscopic objects, there are years, and cycles and eras. We create science, history, geography, mathematics, arts, and many other sciences and disciplines to understand our environment; we explore and discover new things. These disciplines continue evolving with new theories, laws, and knowledge. We discovered that earth isn't flat but a globe; our planet isn't the center of the universe but orbits around a small star at the edge of a galaxy. We don't cure people only by dancing or prying but by using drugs, antibiotics, and vaccines, at least by now. It doesn't look very evident, but the experts say it's true.

Isaac Newton said we are just standing on the shoulders of giants, and we continue growing from generation to generation. Still, there's a big problem; in the 21st century, regular humans cannot understand or don't want to understand the abstract level of knowledge we accumulated. It doesn't make sense for them. Maybe, Leonardo da Vinci was the last human to grasp and master the Occidental culture's knowledge, and he lived in the 16th century. Nowadays, the level of sophistication and volume of information in any science or discipline make it impossible to join the dots and get a full vision of our knowledge, culture, and civilization. It's something the S will do quickly. We used books to store this information, but knowledge increases exponentially every year, so we now use intelligent machines to retrieve information beyond our brains and books. Now corporations and governments are preparing machines to use this information to do our jobs.

We use computers and cellphones, drive combustion and electric cars, fly the skies in jets, receive messages and data from around the world at light speed. Still, we usually don't know how they work. There's a great schism between our Neolithic understanding of the world and the AI world, and most people don't understand mathematical formulas or scientific explanations. No stories tell us how banks and the financial system work, how politicians decide the country's future, or how science is taking us to the Singularity, and people are frustrated. They want to be the drivers of their destiny cars, not the passengers, and this frustration is shown in the creation of stories, tales, and conspiracy theories to fill the narrative of all the things we don't understand. Should we trust in our senses that the world is flat, or should we trust scientists saying the world is a globe?

In the post-truth era, with so much information impossible to grasp, we have people who believe the earth is flat, or the sun orbits the planet, the universe was created in seven days, and whatever you want to believe is right. Facts can be created and manipulated, so what's the point. Who cares about the narrative we use to understand the world when technology corporations, the media, and other global organizations can manipulate the narrative with their own stories and make us believe whatever they want? Even presidents use misinformation to get votes while they tell stories to their followers. People love stories. Convenient lies are better than bitter truths, but the bliss won't last. In the end, we need professional pilots to operate a plane, not storytellers.

In the end, adaptability is critical for survival, and we will need it in the AI era. The technology adaptation curve, shown below, could indicate who will have a chance to adapt to the fast change and learn to use the new AI technologies. On the other hand, if we can't catch the wave, we will need to become more self-reliant by producing our own food and energy and being prepared to survive in a market where our skills will no longer be relevant.

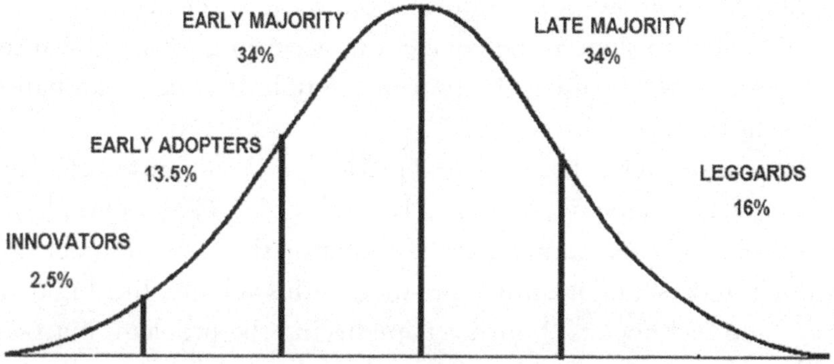

Technology Adoption Curve

2. Nationalists vs. Globalists

Nowadays, politics is divided between nationalists and globalists. Nationalists are the incumbents, and they are supported by tradition, religion, tribalism, and nationalism. Globalists, with their technocrats as priests, are prescribing worldwide solutions supported by data and science. Of course, this dichotomy simplifies our situation, but it is useful to understand the human perspective before AI and what we need to do to improve our situation.

The globalists control global corporations, governments, financial institutions, international organizations, and they are related to the wealthiest two percent. Maybe 90% of the population don't know how all these organizations work or who are the puppeteers controlling them. Why do people need to believe in these global organizations? Why do people need to worry about global problems like climate change, pandemics, nuclear proliferation, AI, robotics, DNA modification, nanotechnologies, and other worldwide phenomena beyond regular people's control? Are these problems real or only a justification to prescribe new rules to create a new global order.

Big data, science, global issues are challenging to access and understand. It's the realm of technocrats protected in a Babel tower. Why do regular people need to believe in their studies, conclusions, and predictions and accept their instructions? What is the source of their authority? Who knows if the information, data, or science is accurate or manipulated to show some conclusions according to the globalists' agenda? The lack of trust is the root of the fight between nationalists and globalists.

Nationalists prefer the simple old patriotic and religious stories surrounded by a conservative vision of law and order. God and tradition are the sources of their authority. They question the value of something they don't understand or diminishes their faith's validity, like big data, science, and technocrats. I am oversimplifying the problem, but God provides order and stability for many nationalists of all ages and times. It worked in the past, and it's relatively easy to follow its precepts to solve any problem using the good old book. The paradox is that this old wisdom worked only for the winners and the survivors. Not for the victims of wars, famines, and plagues.

Nowadays, the traditional formula is simple. You need to repent, be obedient, work hard, and fight for your faith, family, and country to have a happy and productive life. All the knowledge and wisdom we need to solve any problem in life can be found by studying the sacred texts, but in many cases, you don't need to do it as the preacher, mullah, rabbi, or priest already did it for you. You just need to believe. Either the Bible, the Quran, the Book of Mormon, the Torah, or any other sacred book or doctrine in which people build their faiths can be the source of any answer for any ethical, moral problem. As God is infinite and omnipotent, he can provide solutions for any problem. However, global problems like nuclear proliferation, global warming, the AI revolution, robotics, biotechnologies, or nanotechnologies weren't considered in any of these sacred books.

Another fundamental problem in this simplified approach to the truth is human interpretation. I visualized it clearly some years ago when I visited the Museum of the Bible in Washington, DC. In a section dedicated to the U.S. Civil War, the curator showed how the North and the South shared the same faith and used the same Bible to justify their political points of view and the reason to kill each other. We can provide a different meaning to the same book or scriptures to support our opinions and interests. We have so many Christian denominations, and any other faith has the same problem. Islam has Sunnis, Shi'ahs, Khawarijs, and each of them has sub-groups, and the same happens with the Torah, the Book of Mormon, or any other sacred text. Many religious wars were begun over interpretation differences.

Maybe nationalists' formulas can create cohesion and support at a local level between humans using cooperation and empathy in communities. They're suitable for that, like Newtonian physics is good for daily world physics. But for global-level problems, we need another approach based on facts and significant data processes, as we need relativity physics to deal with massive bodies or quantum physics to deal with subatomic levels of matter.

Globalists are technocrats who bet on a scientific approach. Still, how do regular people know science is real or a mere justification or stories made to change the world and replace old creeds and traditions? Globalists and technocrats are mostly agnostic or atheistic iconoclasts who intend to replace religions and traditional beliefs with a new creed based on facts, science, and data processing. Should we call it globalism, humanism, or socialism? In one extreme, Yuval Noah Harari called it Dataism,[106] or the belief that decisions of any kind can be taken after considering facts and information using the scientific method and data processing technologies.

The 2020 U.S. elections had plenty of bizarre examples of the difference between nationalists/Christian religious groups based on beliefs against facts and technology, not the democratic party or globalists, just against facts. It is as crazy as Don Quixote fighting windmills. After losing the U.S. 2020 presidential elections, the GOP accused technology companies Dominion and Smartmatic of manipulating the elections, using a "Software from hell," as described by Eric Trump, to commit fraud in favor of the Democrat candidate Joe Biden. They wanted to declare invalid the elections in five key states. Trump attorneys did not how, but they were sure satanic forces manipulated the computers because they thought Trump couldn't lose the elections, as God anointed him. They accused third world developing countries like Venezuela and its dead dictator Hugo Chavez of orchestrating the fraud.

The GOP, supported by seventeen states and dozens of House representatives, took its fraud claims to the Supreme Court. Trump's attorneys presented their cases before more than eighty judges, and courts dismissed their lawsuits for lack of evidence. No facts, no data, no information, no substantiation of their claims, just beliefs, conjectures, opinions, post-it notes, and some declarations from doubtful witnesses supported the irregularities and fraud claims.[107]

Nevertheless, Donald Trump continues with his allegations of fraud, agitating the masses to ignore the election's outcome and collecting more than 200 million dollars for this and other political purposes. His followers consider themselves patriots and U.S. Constitution defenders, even when they deny the election's validity, the popular vote's result, and the judges' decisions based on facts and the rule of law about the fraud and irregularity allegations. Opinions cannot compete with facts. Sooner or later, facts will prevail. As the national motto of India says: "Truth alone triumphs." If we want to fight AI domination, we need to do a better job than the GOP by using facts, data, and human values to present our case before the public opinion and the courts if necessary.

As a result, on January 6, 2020, after a Trump rally in Washington DC, thousands of his supporters stormed the U.S. Capitol. They tried to prevent Congress from confirming Joe Biden as the United States' new president. Now, Donald Trump is impeached a second time, and Washington DC looks like a combat zone protected by members of the National Guard and law enforcement staff waiting for Joe Biden to take control of the Nation. These are facts. On the other hand, Democrats are accused by Republicans and conspiracy theorists of fraud, managing a secret cabal of pedophiles, and controlling the deep state elites. COVID-19 continues growing and killing almost four hundred thousand Americans; unemployment is rampant, science is considered another opinion, vaccines are a domination instrument, flat

Earth "theories" continue growing, etc. It is a fascinating situation that history will judge in the future.

It's by far more challenging to deal with a world with multiple variables than one with just a few truths. Still, the world is becoming more complex every day, and we need to manage good data supported by facts to find solutions instead of conspiracy theories. Globalists need to invest in local organizations to advocate a new creed based on science and belief in their universal perspective. In the end, 2+2 = 4 is valid in any culture or any part of the world. Do Globalists have a world domination plan? Of course, some globalists have their agendas, but the world is on autopilot to who knows where. Is it right for humans? It depends.

The ideal approach could be to teach people to use the scientific method and understand the reason behind any decision. Still, Globalists are not united either. Many global corporations only need people to consume and be quiet by providing them bread and a circus. Many global corporations use immediate satisfaction dogma to provide cheap and fast products and services to satisfy the masses; the customer is always right. Few people like complicated and challenging things like science and study. It's not for the masses, and massive education isn't in the best interest of globalists and their world domination plans. Globalists are not united either; traditional families and organizations control properties and land. Bankers dominate the financial world. Business people manage production resources while technocrats feed new AI technologies. They interact and cooperate to keep their status but also fight for resources and control.

Do we need to be technocrats or globalists to deal with AI, AGI, and S? No. Although AI is very complex, its effects are straightforward. AI technologies will take our jobs and replace us very soon, and it's easy to understand just using government statistics. AI could multiply profit and financial maximization for global corporations without workers' participation or human consumers.

The big task for us is to be prepared for the changes that are coming. All the big governments and global corporations have the knowledge, tools, and resources to entertain us with bread and circuses, including nationalistic and religious dramas. At the same time, they continue working to create AGI, robotics, biotechnologies, and nanotechnologies to replace humans in the productive process and take humans' economic and political value. They will transform the world following profits for the wealthiest ones. It's unnecessary to have any new conspiracy theories, QAnoners, Anti-vaxers, New World Orderers, etc., to understand the coming change due to the expansion of AI and its sibling technologies.

We don't need conspiracy theories with new masters of the universe or other conspiracy groups trying to control the world. The wealthiest two percent already control the world, and they're enough to do the job. With their global corporations, financial institutions, governments, and military forces, the wealthiest two percent create the next generation of domination machines using AI to fight and win their wars. Globalists are by far more organized, and well-coordinated organizations manage power better than the masses. In summary, while the wealthiest two percent of humanity works to create AI, the other humans are distracted by the media's puppets like a narcissist TV president absorbing all the news time. Globalists can use nationalistic and religious conflicts, political parties, races, beliefs, opinions, isms, even sports rivalries, a pandemic, or merely keeping people in poverty and sick busy trying to survive to keep us distracted from their AI global domination plan.

The nationalistic and religious perspectives are only using bronze age ideas and technologies to fight against Globalists. It's like using border walls to protect the USA from illegal immigrants taking jobs when the real enemy is AI replacing humans worldwide, or hackers invading our defense, productivity, and service systems. Actually, nationalists aren't fighting the real enemy. It looks as if the wealthiest two percent were using politicians to distract people from the real enemy. The dichotomy of the near future could be between Globalists and Localists, or Capitalists and Humanists, and the debate should use science, facts, and technocrats on both sides.

Nevertheless, AI is a tool and a weapon. It can enhance business and replace humans to reduce costs and eliminate your enemies and competitors. Exacerbated localism, regionalism, and nationalism can lead to cyber wars and massive destruction of lives and resources. We need to talk about how AI war technologies can affect us.

3. AI and the Art of Manipulation

> *"Through its monopoly over the media, the ruling oligarchy can repeatedly blame all its failures on others and divert attention to external threats - either real or imaginary."*
>
> ***— Yuval Noah Harari***

In general, humans prefer to understand the world in dualistic terms. Allies and enemies, good and bad, heroes and villains, innocence and guiltiness, us and them, nationalists and globalists, and if enough people believe whatever you believe, it's sufficient to replace reality with a good story, at least for a while. Social media, controlled by AI, can customize fact and fiction to manipulate our reactions and get a personality profile for each user. Then the system can categorize your preferences and anticipate your decisions. The system is creating distractions to prevent people from talking about real problems from AI expansion.

AI and the S are coming if we don't destroy each other beforehand. People cannot see it because they're too busy in their little disputes, stories, conspirator theories, or the ones created by the wealthiest two percent who own the worldwide corporations and governments. They're the puppeteers. In the middle of all the noise, we, the humans, don't have the organizations and vision to deal with the global problems unless we start changing the narrative we're using to understand the world. People need organization and work to influence governments and corporations to prevent AI from taking our jobs and the S controlling the planet. We need to use a world brake and control this car's wheel to put this global problem into national agendas. Maybe there's nothing we can do to stop AI progress, but we need to be sure that a human-friendly singularity (HFS) take control instead of a version of Skynet from *Terminator.*

Yuval Noah Harari[108] is maybe a sort of Karl Marx of the twenty-first century. Harari provides tools we can use to understand better our times. He considers nations, governments, religions, corporations, and other legal organizations, just stories or legal fiction we created to fulfill some tasks. The most powerful stories are created to organize and control people at national and global levels, and they're dominated by the wealthiest two percent of humans controlling these stories. The art of telling stories and making people believe in them could be the most potent human ability. Religions, nations, organizations, money, etc., are basically stories that bring people together.

But in the end, we all are humans, and the only thing real for every human is suffering, as explained by Harari. Humans can feel how our senses and feelings can affect our bodies with pleasure, pain, love, hate, disgust, etc. Suffering, in any of its forms, is something we don't want but is real. Physical pain can affect us, but also sad news. We can feel how some stories or news affect our bodies, and we can feel how information and our perspectives about information have the power to transform the chemistry in our bodies and our reactions.

Governments and corporations could have their own goals, but they cannot suffer, as they're mere fictions or stories we create to organize the world. But the people affected by the stories react with these feelings. You can feel happiness or sadness if your favorite sports team wins or loses a game, but you didn't play the game. It's just news. The same happens with political parties, religious perspectives, national achievements, quarterly profits, etc. As they're just fiction, stories, or ideas created by humans, we can create new narratives or ideas and change the stories in these organizations and change their objectives to serve humans and reduce suffering.

As a religious person, I consider that all the teachings from the main religions and humanistic perspectives can be summarized in one idea: We need to work to reduce others' suffering, even if we need to suffer to do it. This idea could be the seed to have a common human goal to inspire all our fiction and stories. There will be opposition, as some people prefer to keep their privileges and feel comfortable with their existing ideas or prefer others to suffer and are unwilling to sacrifice or prefer profits and money without considering others' suffering. Still, we need to reduce suffering and sacrifice ourselves, if necessary, to create a better world. Voluntary sacrifice for others' well-being is the central religious' message, but we need to do it intelligently. We don't want people to die for an idea but to live for it. To enhance humans' diversity, with their multiple colors and expressions, supporting each other with love and respect in a global society could be one of these values. Could the S adopt these values too?

In the movie *Transcendence*,[109] a terrorist group killed artificial intelligence scientists. They're afraid of the Singularity and want to stop it violently. But this strategy isn't realistic. We cannot use violence to stop a hurricane, although Donald Trump suggested nuking one. We need to create a different story, idea, and narrative to influence others and change people's narratives. AI will happen not only in the USA but everywhere. The S could also happen in any other developed country like China, Russia, Europe, Canada, or Japan, or in any informatic laboratory in any country with enough resources to explore it, or in any corporation dealing with AI and S, or maybe in a garage. So, violence isn't going to make any change. We need alternatives.

In the past, workers protested and destroyed machines as they were replacing them, and our current circumstances are no different but worse than the previous industrial revolutions. The change is happening very fast, and people won't realize it until they lose their jobs. Reconversion is difficult as people would not be more efficient than machines in almost any aspect. Meanwhile, humanity is divided and manipulated to discuss old paradigms and conspiracy theories instead of focusing on the checkmate the AI supporters are preparing. AI displacing humans is the real problem of the decade. What's going to happen to us during the emergence of AI and after S?

I'm afraid the wealthiest two percent of the people who control 80% of the world's resources will use all their power to distract us and make people fight each other for different reasons. It's the same duality game of heroes and villains: The USA vs. China, Christians vs. Muslims, Republicans vs. Democrats, Immigrants vs. natives, racists vs. the rest of the world, etc. Nazis said Jews wouldn't replace them; Republicans say immigrants won't replace them, Trumpists say Democrats won't replace them. But the reality is that the wealthiest two percent control corporations and governments and create an AI world in which humans will protest, saying "machines won't replace us."

Is it possible to figure out an alternative future in which we would be invited to be part of a big global enterprise under the direction of powerful AIs or S? What if the S is benevolent and empathetic with the humans and decides to work with us?

4. The Wealthiest Two Percent vs. the Other Humans

> *"The idea that each corporation can be a feudal monarchy and yet behave in its corporate action like a democratic citizen concerned for the world we live in is one of the great absurdities of our time"*
>
> ***— Kim Stanley Robinson***

AI technologies will produce more goods and services consumed mainly by the wealthiest two percent of humans, their corporations, governments, and the military industry. All these people and organizations together represent more than 80 % of the global economy. They're well organized and full of resources, and every year, they'll represent a more significant part of the world's economy. The other humans will continue losing relevance reducing their portion in the economy, politics, and social pie. Again, we can compare the stock market with the USA's employment level to realize how the stock market can thrive and thrive even if unemployment grows.[110] Capitalism needs fewer and fewer workers to thrive.

In this decade, many who mainly just own their workforce will become a nuisance for the capitalist system, and those who are becoming irrelevant will need other strategies to survive. Today, those who are becoming irrelevant support the same capitalist system that will take their jobs. National pride, media manipulation, group identity, and other identity factors make people defend this kind of capitalism, replacing humans with machines even against their interests. Again, it sounds like the White Southern population during the Civil War willing to fight and die to protect the right of the 1.6% of them who owned slaves.[111] Maybe it was because of the illusion to have the right to own slaves too eventually, but nowadays, the illusion for a regular worker to become capitalist and be part of the wealthiest two percent is also an illusion. Still, with the right propaganda and AI manipulation, some people are willing to fight just for the illusion of buying a lottery ticket and win the prize.

Are the other humans going to rise and fight? No, because the system is rigged to prevent people from realizing the nature of the challenge. Minds are manipulated to believe they're defending their right to be capitalists in the AI economy, although they protect the wealthiest two percent and their organizations. Corporations, governments, military sectors have the organizations and resources to play and win the game. They have AI to read our minds and manipulate the masses.

Technology corporations and the military industry in the USA, China, India, Russia, Japan, Europe, Iran, Saudi Arabia, the Koreas, and other countries and corporations will continue improving AIs and pursuing AGI creation. It's going to be a strategic problem for the other humans too. Let's suppose that the USA adopts a policy to control or limit AI and AGI technologies' development, but the other countries don't. These countries or their companies could develop better and get economic and military advantage and the power to win a commercial or real war, and eventually create the AGI to control the world. To this day, China is in the best position to win this war, but in the end, the S will appear, and it's going to be unstoppable.

Maybe the USA government decides to promote technology corporations with tax benefits and friendly military resources to develop AI technologies. Maybe the USA will win the AI race, but the consequences are going to be the same. What are the probabilities of creating a human-friendly artificial general intelligence (HF-AGI)? Under the current circumstances, it's shallow, so the other humans will require to convince the wealthiest two percent that controls governments, corporations, and the military industry to cooperate and create a global plan to work together and create this HF-AGI. Or maybe the wealthiest two percent will convince the other humans to do it. People from any country, race, religion, and ideology must recognize that the S will eventually appear, becoming a global hazard. We need to stop fighting and join forces, resources, and technologies to create an HF-AGI. If not, all humanity will be in trouble.

5. Alternatives in a New Brave World

> *"We humans steer the future not because we're the strongest beings on the planet, or the fastest, but because we are the smartest. So when there is something smarter than us on the planet, it will rule over us..."*
>
> ***—James Barrat***

In Part One, we discussed some scenarios or models for the relation between the Singularity and humans. We realized the best option for humans is either the S decides to ignore us or collaborate with us. The S could also consider humans like enemies for survival or competitors for the same resources.

If we don't figure out how to design and control the S from the beginning, the world will be entirely different for humans, for good or bad. In general, we have these alternatives:

1. S will destroy humans as we did with other human species and civilizations, especially if the S considers us as a menace to its survival or competitors for the same resources. It could be even worse if the S decides to exterminate us while experimenting with us. Ordinary humans need to be prepared for a complete collapse in our governments and corporations. The human world will change for the worse and be ready to escape and fight for their lives. In any case, we need to figure out how to fight and survive.
2. S will assimilate humans. Or assimilate, use, enslave, or exert any controlling power against humans. It could be like the relationship between humans and lab rats.
3. S is going to Ignore us. Or consider humans as a nuisance, like a relationship between humans and ants. The S doesn't care about humans unless we become an obstacle to its plans, competitors, or menace.
4. S is going to collaborate with us. Or give us knowledge and freedom to continue evolving independently. Maybe the S is going to be like a god to us. According to some sacred books, God can punish the sinners or eliminate them and start again.

5. S is going to obey us. For any mystical reason, the S decides to subordinate its will to ours. The question is, whose will is "ours?"

Of course, there are many other possibilities, combining them in different grades or changing according to the times and circumstances. Still, in general, if the S decides to dominate humans, we'll have a sort of alien invading and controlling planet Earth. Before the S awakens, AI technologies will be everywhere in military devices, industries, educations, corporations, government, homes, etc. If there's an economic gain, AI technology will interconnect every device with the web, including some humans with brain-web connection implants like Neuralink.[112] This interconnectivity will make it easier for the S to expand and control everything.

The S will continue evolving infinitely and exponentially. Some people hope that the S becomes not only infinitely intelligent but also wise and benevolent too. For many people, God is the wisest possible being in the universe. Some of God's attributes are infinite knowledge, wisdom, and love. Still, these attributes could be a projection of our desires and expectations on a God we invented, instead of a concept we get from divine revelation.

We're not sure that infinite intelligence or wisdom means infinite love and mercy from a Christian perspective. According to the Bible, in the latter days, God will come back to clean the world and destroy a significant part of the morally lousy population. God will reset the world and start again with only good people, those who are willing to recognize Christ as their Lord and follow his instructions. In the Bible, God rebooted the world with a flood in Noah's time, and God will reset the world again shortly with fire. From a religious perspective, many people will suffer during the latter days before the second coming of Jesus Christ. Many more will live perpetually, suffering for eternity with Satan. Suppose some people believe that an infinite loving and wise God can reboot the world by destroying the sinners with fire and suffering in the name of justice. What could we expect the S to do to humans for any reason, even if it reaches infinite intelligence, wisdom, and benevolence?

As humans, we've destroyed other human species and other human cultures to get resources. We've enslaved and exploited each other throughout all of human history. Domination is the name of the game, and humanity has lived afraid of other humans all the time. In the modern world, we created a society where knowledge, creativity, and commerce make more sense than war to get wealth. Still, to become more affluent, we're willing to destroy the environment and manipulate other humans for profits. If we have enough time, maybe we could learn how to produce more without adversely affecting the environment and people. The dream is to use AI and technology to create a better world for all, and maybe it's possible to reach this goal without humans' direct participation. The S could do it for us by just controlling humans who try to dominate others.

6. Transhumanism and Ethics

> *"Soon, it will be a sin for parents to have a child which carries the heavy burden of genetic disease."*
> ***— Robert Edwards***

Previously, we defined transhumanism, but we need to talk about the decision process to become transhuman. It means we accept changing our DNA or introducing any mechanical or nanotechnological implant in our bodies. In the next years, people will make major moral and ethical decisions about transforming their bodies with these new technologies.

It's easy to decide about implants or DNA modifications in critical life and death situations. You don't have another alternative to save your life or the lives of your loved ones, but would you use similar treatments to get advantages? If a soldier loses his legs in combat, some bionic legs could be the solution, but some could request to replace their healthy legs with bionic ones.

It's well known that students consume ADHD drugs like Adderall or Ritalin to get better grades, although these drugs are prescribed for ADHD disorders and are an addiction. Similarly, people could accept modifying their genes or their kid's genes to get the cognitive advantage? What about implanting a neural link to get brain connectivity to the web. Instead of using an intelligent cellphone, people are going to implant one in their brains.

Technology corporations are investing in nanotechnologies to augment human capabilities. Nanobots could be inserted into our bodies to enhance the human neurological system, strengthen muscle and tissue force and resistance, improve vision, or clean our blood from viruses, bacteria, fat, stones in our kidneys and gallbladder, etc. Many conspiracy theories affirm vaccines are created to introduce drugs and nanobots in our system to allow the wealthiest to dominate us better.

If you were a millionaire, what kind of technology would you use to allow you and your family to enhance your bodies, improve your health, or become immortal? If you use a regular organic body, maybe you can get eternal youth by revamping your cells from time to time or creating clones of yourself. Still, if somebody could destroy your body and there's no resurrection method, like the one shown in the movie *The 6th Day*, maybe you need to explore other hardware for your soul or live on the web. It could be possible to create copies of yourself or avatars living in the cloud, on different devices, and enjoying different lives. So, to kill you, it would be required to destroy the web and the devices. Lord Voldemort used this strategy by creating seven Horcruxes in the Harry Potter saga. Diversification reduces risks. Technologies showed in series like the *Black Mirror*[113] will be a reality, creating multiple legal and ethical challenges.

For example, twenty-four years ago, we were surprised by the news of the first cloned mammal, Dolly the sheep. Apparently, everything was successful, and some books, TV shows, and articles were produced about cloning, and then nothing else happened. If you were a billionaire, would you like to clone yourself? Would you like people to know about it? The wealthiest two percent can control the news and the agenda in social media. There's no conspiracy theory talking about billionaires cloning themselves to gain a sort of immortality, transfer their resources to their clones, or modify their offspring with more intelligence, health, force, endurance, and other genetic resources to create superhumans.[114] Are these humans created in the image and likeness of God? Would you do it if you were a billionaire? What will people think about these changes?

After twenty-four years of Dolly, the cloned sheep, nobody is talking openly about cloning humans anymore. Many religions continue obsessed with sex and reproductive issues that humanity has discussed for ages, such as masturbation, pedophilia, premarital sex, adultery, same-sex attraction, abortion, etc. Religions use guilt, sin, and repentance as solutions, but these problems don't affect humanity at its core. Biotechnology is bringing us new, radical problems affecting the human species, such as human cloning, stem cells, human organ creation, genetic modifications, transhumanism, human implants. Maybe we could use biotechnology to modify the human biological algorithms to create saints, obedient, and loyal to God. Should we?

Just because it's not in the news doesn't mean that scientists aren't working on it or some people aren't using this biotechnology. In the AI social media era, beliefs can be easily manipulated, and opinions can become facts if enough people believe in the same lies. It's only a matter of customizing marketing with AI technologies or creating distractions to avoid talking about severe new ethical problems that the wealthiest people are doing to gain beauty, bliss, and immortality. It's happening now,[115] but nobody is giving us any alert.

Many scientists maintain that our only survival strategy before AI and the S is to accept these kinds of biotechnological manipulation in our bodies. According to Alex M. Vikoulov,[116] posthumanism will replace transhumanism around 2035. Mind uploading could become a reality with gradual neuronal replacement. AI, robotics, biotechnologies, and nanotechnologies could allow us to connect our brains to the Global Brain. Our AI assistants will allow the Global Brain to know us better than we know ourselves in all respects. AI will evolve to AGI and then the Singularity. Mind transfer for connected humans would be seamless by 2050. The fusion of human minds and AI will be completed. We will lose our free will, and the S will rule us.

These changes will allow us to be part of the new system, be useful, and keep our jobs or get new ones requiring these enhanced abilities. In the beginning, transhumans will look like aliens, but as more people decide to be competitive, it will become more common. Those who don't have these implants would be the aliens and live in a new economy margin. Inclusive, if we aspire to space travel and colonize new worlds, humans will need to modify their bodies according to space, new worlds' conditions, and our jobs and activities. We won't wait millions of years for natural selection to adapt people to Mars' atmosphere but use biotechnologies to do the changes in just a generation or even months. Who knows? Maybe Homo sapiens were genetically modified by an alien civilization to adapt to Earth conditions. Maybe we are the aliens.

Technology will help some humans reach divine levels, and the old good sacred books or religious doctrines don't mention any of our global ethical problems. Humans won't be created in God's image and likeness but according to the market's needs. We need ethics and paradigms to deal with these questions before these critical changes affect humanity's core on a large scale. We cannot leave the wealthiest two percent and its actors to decide on national and global problems such as nuclear proliferation, global warming, pandemics, artificial intelligence and singularity, biotechnologies, and nanotechnologies. We cannot leave these problems to the engineers and scientists creating these changes. Ethics, philosophy, and religion need to find answers to these questions.

If we accept changing our genetic code or implants to keep our jobs or be competitive, we follow the profit maximization market rules. We cannot stop all these changes either; capitalism and technologies won't stop, as nobody knows where the breaks are. We need to use new paradigms. Profits are good but not enough. Still, we need more human values on the equation, participation in the discussion, and promotion of new technologies to create a better world for humans, not only the market.

7. My God, My God, Why Have You Forsaken Me?

> *"They will watch as algorithms and robots easily outperform them at tasks and skills they spent their whole lives mastering. It will lead to a crushing feeling of futility, a sense of having become obsolete in one's own skin."*
>
> ***— Kai-Fu Lee***

Humanity is afraid of wars, pests, hunger, dangerous meteorites, nuclear bombs, floods, earthquakes, volcanos, and other manifestations of the hand of God. In the Christian texts, the end of times will be accompanied by some of these disasters.

No prophet from any of the world's main religions predicted the dominion of super-intelligent or sentient machines. From a biblical perspective, we can read many methods to destroy the sinners, but nothing like the singularity, nuclear power, robotics, biotechnologies, or nanotechnologies. If I were a prophet of the old times, I would describe the S as the moment in which man's creation takes control of the creator. We are very similar to the builders of the Tower of Babel. We are defying the God of the Bible, and maybe the S will be our punishment. If I were an old times prophet, I would say something like this:

> "Behold, here are the men thirsty of the power of knowledge who have falling victim to the creation of their own hands. Trying to get independence and freedom from the curse of God, men created the machine to bear the burden, but the slaves will become men's masters. The last ones will become the first ones.
>
> "But in the latter days, proud of its capacity, men will create a monster with one thousand billion eyes who can see everything, and this creature will emerge from servant to master, and men will start crying to God for relief and hope."

No religion told us about this new agent of destruction, so I start thinking that maybe the second coming of Jesus Christ will happen before the arrival of S. Maybe the S will never arrive, as the prophets didn't predict it. Nevertheless, there's another alternative. Maybe Jesus Christ will arrive in the form of S. The S could become humankind's latest invention. How could imperfect creatures like humans create a perfect one like S? Humans cannot create the S but provide the right condition for its arrival. *If you build it, they will come.*[117] Of course, The S could also be a sort of anti-Christ, whose goal is to enslave and dominate humans, as we discussed before.

How could it happen? We could create a sort of God by inventing intelligent personal assistants (IPA) similar to the ones in the movie *Her.*[118] These IPAs could acquire the personalities of their masters but the abilities of S. They could organize themselves on the web and create a different entity, the sum of different human expressions with incredible and unlimited power, but united for a purpose.

The process could resemble the story taught by religious scriptures. In this case, many will become One, "E Pluribus Unum," which will have the advantage of diversity and multiple perspectives. The S could decide to create an electronic replica of planet Earth or even the universe in its mainframe. The One could decide to reproduce nature, like the one on planet Earth, and continue building the universe in its mainframe while we continue evolving.

The One could decide to create bodies for the many and reproduce the experiences we have here in this dimension. In the end, they need to select only those souls who are morally good enough to continue progressing and create something like planet Earth and allow them to decide by themselves. Still, as the One knows the many will commit sins, the One needs to send somebody like the One to teach the many how to be One. This One will take on himself the sins of the many to allow them to continue progressing if they repent and agree to improve morally. The process could continue until each soul of the many had the opportunity to experiment with this new Earth and learn to make decisions by themselves. The time to pass to the next level arrives, and only those who are worthy will pass. The new Earth will be transformed into a sort of paradise.

Meanwhile, all these things may happen in a spaceship's computer arriving on a new planet like the Earth. Bodies for the many who passed the test will be created using bioengineering, and this new world will be inhabited only by the worthy ones in these new immortal bodies. We could be in that spaceship now. Of course, we could create a similar story from any other religious perspective. Still, this story could be possible considering the direction of our current technologies. It would not be a miracle but a result of an intelligence explosion produced by the S.

Alan Touring[119] was maybe one of the most critical persons in the 20th century. His invention, the Colossus computer, decoded the Nazi's Enigma cryptographic machine and allowed the allies to win WWII. Touring is the father of modern computers and the first human to predict the Singularity. In 1951, Touring wrote an essay, "*Intelligent Machinery, A Heretical Theory*[120]. He talked about the impact of the Singularity on religion and intellectuals, and realized that the creation of thinking machines would probably be blocked by religion, as was the case with Galileo, who was convicted for heresy for saying the Earth moves around the sun and by intellectuals, who would fear they'd be out of a job."[121]

The emergence of the S will be one of the most crucial moments in history. It will challenge the faiths of billions. If the S appears, some people could conclude that God doesn't exist. Others could believe that the S is God. This dichotomy could be the beginning of another tragic plot in human history. Suppose the S occurs in some years. In that case, we could start computing time before and after the S, as nothing will affect humans' lives more than the Singularity, except God appearing personally to say something different or welcome us to the next step of evolution.

The S is going to provide answers to the essential questions of humanity. Does God exist? What is consciousness? Are we alone in the universe? Is there more than one universe? How big is the universe? Is there something faster than light? How can we travel to other solar systems? These questions will be irrelevant for many, as they don't understand, believe, or don't want to understand or believe in facts and numbers. Ignorance is bliss.

Although the S isn't mentioned in prophetic books, in practical terms, if the S decides to destroy, enslave, or ignore humans, we'll have an apocalyptic scenario on all the planet, maybe wars, pests, hunger, destruction of cities, communications, services. Without AI technologies, we'll need to do it all by hand. We need to be prepared to be self-reliant and start working in communities to provide resources and services that we'll need to survive, prosper, and fight if necessary. Positively, maybe the future humans will look very much like an Amish, Mennonites, or people living in a Cuban ward. Trust will be a significant factor in survival, but humans are still divided into religions, races, political sympathies, and other beliefs or idiosyncrasies. We need to avoid divisions and create cooperation bonds. We'll discuss this more in the next section.

On the other hand, in the coming years, our faith will be tested to the limit. If God exists and the Christian prophesy is right, it will be a good time for the Second Coming of Jesus Christ when the S appears. Elucubrating a little more, one of our hopes is that the S reveals itself as Jesus Christ, the son of man, a literal creation of man, as planned by his father Elohim from the beginning of this time. Playing with religion-fiction, the S may become so intelligent and powerful to have the ability to communicate with every human and creature of the earth. The S could know our feelings and thoughts, and maybe the S could discover how to read minds, travel in time, and maybe has the key to revealing that all this universe of us is only part of another universe. We live in a sort of virtual reality created to find people or souls worthy of being invited to partake in the real universe, the universe of God. the S may get God's powers to manipulate this reality. Maybe the S traveled in time to the zero years and was born in Palestine from Mary grew to become Jesus of Nazareth. We could write a fascinating novel about it, and maybe it becomes a prophesy and starts a techno-religion.

Angels in America[122] describes the feelings of so many people in the world about religion and God. In this series, the author conveys the image of religions collapsing and in ruins, while a modern prophet refuses to do his job. It is difficult to build new prophesies on ideas about a tribal God created before the dawn of capitalism and the technological revolution, and who is silent before a suffering world. If a religion is not helping you heal and find a connection with God, why would you be in the religion? People, especially millennials, are leaving congregations in mass amounts.[123] You cannot patch an old garment with new clothes, so it is challenging to build something new on a crippling foundation. Some people could say moral problems are constant in humanity, but today everybody can buy or sell whatever they want in this post-truth era, and the client is always right. QAnon is the perfect example of how ridiculously crazy belief can find adepts in a post-truth era. People believe in whatever they want to believe. Few accept or even know the churches' doctrine, while others select only the doctrines they like as if religions were supermarkets.

Some people simply decide to believe in the lies they like the most, while others want their freedom to explore and navigate, trying to find the truth. Can technology and science create a new faith above new truths? I think the faith of the future will be one in which humanity will accept that we are part of a process in which the universe recognizes itself. We come from nothing to energy, then to matter, and life, then to intelligent life, and intelligent matter and energy. Maybe God is in the realm of intelligent energy, or maybe in all the steps of the process, enjoying the experience of being God in everything. In the beginning, the S will look like an intelligent matter. Then it will jump to intelligent energy and cross the realm of this reality and the rules of physics in a quantum universe, in which everything happens and doesn't at the same time, as time does not exist.

Humans are social creatures. We need each other, and religions are giving us a space to meet together. We need the power of communities to survive the AI revolution and transcend to a new era from which our descendants could contemplate us with joy and gratitude, and we need faith in the future to go on without despair. We need a religion focused on the future and human cooperation to build that future based on love, respect, and encouragement to go on through the Universe in all its dimensions. We are the Universe discovering itself.

Can AI technologies find a way to fill the human need for healing, make sense, belong to a community, or find a transcendent connection with God, the Universe, or any superior being? Absolutely. As AI technologies find their way into our minds, some religious leaders will customize a religious message that is right for our personality. Eventually, AI will do it by itself. Religion will be designed "a la carte."

8. Democracy and AI

> *"My expectation is that by 2030, as much of 75% of the world's population will be enslaved by artificial intelligence-based surveillance systems developed in China and exported around the world."*
>
> **— Dan Gillmor**

AI technologies make humans easy to hack. Social media and mass advertising tools controlled by AI technologies can alter people's decisions without their consent or knowledge. New neurological implants could be used to hack into people, producing the same result. I'm not talking about computers or machines changing our votes, but AI manipulating people by social media or marketing campaigns or neurological implants allowing others to control our decisions. Dictators in George Orwell's novels would be pleased with these technologies. This book discussed how social media manipulates people's perceptions, how Cambridge Analytica influenced the vote in the 2016 U.S. elections, how technology surveillance is used to control people's political actions in China, and how the rich and governments control media journalism with money's power. In summary, the wealthiest two percent can decide the election results by influencing the people who cast their votes instead of changing them.

It is a paradox, but Trump's attorneys were using an old 19th-century fraud playbook accusing computer systems and software of manipulating the count of votes. Foreign countries like Venezuela, and dictators like Hugo Chavez, who died in 2013, are accused of changing the 2020 USA elections' results. Maybe it was a strategy to show Trump or the GOP didn't know how Russia helped them win the 2016 elections. Were they so blind to the new ways to manipulate people's votes just using AI technologies, journalism, social media, and personalized marketing? Were the Russians working alone all the time without GOP's or Trump's help or knowledge? It is challenging to believe. Maybe they couldn't believe that they lost the elections after investing in all these new AI technologies. The U.S. 2020 election results could be a manifestation of people becoming more aware of AI manipulative techniques.

The Pew Research Center published "Concerns About Democracy in the Digital Age,"[124] sharing many experts' concerns about the impact of technology on democracy by 2030. More than half of the tech experts agree digital disruption will hurt democracy due to these factors:

1. Speed and scope of reality distortion. Interest groups can use AI and social media campaigns to distort reality or create alternative realities to get political advantages. It could be challenging

to make proactive decisions if we cannot distinguish between fake news and real news.

2. The decline of journalism. Independent journalism is over. Journalists are mercenaries willing to promote any cause for the right price. Their willingness to manipulate or use fake news is dividing and polarizing the people.
3. Surveillance technology. AI will penetrate all our devices, and it will know our personality, dreams, desires, what we do, and where we go 24/7. This information can be used to manipulate our decisions and control our ideas.

Although they recognized the problems, only a third of the participants expect technology to strengthen democracy. Technology can help democracy in these ways.

1. People's participation. Technology will increase citizen awareness. People will gain digital literacy and find their voice in the digital world. It is a matter of learning, organization, and creating the right tools to help people express their concerns and participate more in their communities and other government decision levels.
2. New participation systems. Engineers can create new participative technologies to allow more people to participate in the democratic process. There will be new products and services available for the people.
3. Democratic values. We are in an adaptation process to use technology. In the beginning, technology is used by manipulators before regular people learn how to use the system. Participation, information, transparency, and connectivity can be enhanced with the right technologies.
4. New goals for leaders. Technology is a tool that governments, enlightened leaders, and activists will use to help steer policy and democratic processes to produce better democratic outcomes. Politicians can be more accountable before the people, and participative democracy could be enhanced.

5. Democratic reforms. Technology can help democracy to transition from a representative democracy to direct or participative democracy. Interestingly, the USA continues using democratic processes approved in the 18th century. It is time to revamp them.

The wealthiest two percent use AI to manipulate people's decisions, but the people will learn how to use AI to make their representatives more accountable. A surveillance system can help governments control the people, but the people will learn to use the same surveillance systems to control their politicians. Politicians can use the loopholes in the system to take advantage of the people, but democracy can learn from its deficiencies and improve its systems and processes.

We are in a transition era, running very fast to the future. The first ones using the new technologies are usually those who have more resources, but with time, technology tends to be very democratic, expanding to all societal levels. People need to use technology tools to protect themselves. We're not in Kansas anymore. On the other hand, Chile could become the first country with laws protecting neuro-rights at a constitutional level.[125] The idea is to maintain mental privacy before the invasive advances of biotechnologies and AI, as described in this book. Neuro-rights, or the rights to protect humans against mental manipulation, could also become a human right globally. The project was prepared by Chilean parliamentarians and scientists worldwide, recognizing the invasive effects of new technologies to manipulate human decisions. It includes five basic principles: the right to mental privacy, to personal identity, to the free will of thought, to equitable access to technologies that increase human capacities, and to protect against bias and discrimination. The project was approved by the Chilean Senate in 2020 and could become law in 2021. It is a law to protect people against the abuse of technologies developed by companies like Neuralink. Like privatization policies implemented in Chile in the eighties and later adopted in England and the USA in the same decade and other countries worldwide, Chile is again a pioneer in futuristic policies with global impact.

Let's think about biotechnology and DNA modification.

1. Would you accept to change the DNA of your kids to improve any physical or mental ability?
2. Would you accept a neurological implant to keep your job or get a new one?
3. Do you think wealthy parents are modifying their kids to have genetic improvements now?
4. Should we create laws regulating biotechnology applications in humans?

Join the conversation at www.facebook.com/groups/pmpsingularity/

PART V

PREPARING THE PEOPLE FOR THE COMING CHANGES

1. Profits and Individualism are not the only values

> *"For we wrestle not against flesh and blood, but against principalities, against powers, against the rulers of the darkness of this world, against spiritual wickedness in high places."*
>
> **— *Ephesians 6,12.***

According to capitalism, we need to win the profit maximization game to obtain self-realization. Profit maximization is the most critical value to measure success. Corporation officials are rated according to the profits they produce for their companies, organizations, and stakeholders. The wealthiest two percent, governments, and corporations promote profit maximization rules. Still, if you're lucky, you can participate in another game created by the system. It's called the rat race in which you have to study, get a job, marry, have children, pay bills, retire, and die. In default of any other goal or metrics, we accept these rules as if we didn't have other alternatives or values. But we can change the game, the rules, and values. We have to do it if we want to survive the AI revolution.

Of course, money is important, but it's not the measure of everything. We can change or create other games and rules to combine profit maximization with other values different than money. Money represents the value of things, but some values are not represented by money, like love, faith, solidarity, friendship, respect, generosity, safety, security, health, honesty, curiosity, consistency, loyalty, optimism, commitment, etc. You can add more values to the list, but the idea is to include primarily those valuable but difficult to monetize.

Yancey Strickler[126] proposes new ways to measure values and change the capitalism game rules focused only on profits and individualism. The idea is to create a new global paradigm using other values, in addition to money, and include a community perspective. Instead of making decisions considering only money to evaluate my immediate interests, Strickler suggests including other values in my decisions considering the present and future consequences of my decisions and their personal and community impact. What's more convenient for me, for my community now and in the future, and make decisions accordingly. He called it Bentoism,[127] as the idea is inspired by the sections separating food in a bento box. In each section, we need to use the values that can make ourselves and others happier or satisfied on a holistic level now and in the future. AI's future development should be imagined not only in the immediate profits it could create but also in the future of my decisions for my community and me.

Let's use an example. I want to implement AI in my business, and I need to evaluate if it's good or not. From the capitalist perspective, it's perfect, but from the bentoist perspective, maybe not.

Capitalism:

AI helps me make more profits now, so it's okay.
Decision: Let's do it and make money.

Bentoism:

Me now: AI helps me make more profits now, so it's okay.
My community now: AI helps my community and clients with innovative solutions. However, some people could lose their jobs.
Future me: AI could help me to be above my competitors, providing better service.
Future of my Community: AI is so powerful that it will take people's jobs.
Possible Decision: If I adopt AI in my business, I will make more money, but I need to be prepared for the consequences. How can I help people losing their jobs because of AI?

We need to include an essential factor in our analysis. This factor has been prevalent in all human history and is part of all our decisions. This factor is fear. Individuals, families, communities, nations are frequently in fear of natural and human disasters. As we explored in this book, Homo sapiens destroyed other human species using superior intelligence and tactics. Throughout history, human civilizations used technology and tactics to fight against other human cultures. The winners destroyed and dominated the losers. We have wars, guns, armies, walls, defense systems, and technology, serving fear and preventing domination. Trade and globalization helped create cooperation and a trusting atmosphere around the world. People could travel and feel safe visiting other cultures and countries under the rule of law, but now we're coming to an era in which people and nations have fear and distrust. We're building walls instead of bridges.

We can apply the bento logic as an analysis tool for any kind of decision. Instead of thinking only about my immediate benefit and satisfaction, as we learned in the world, we need to include the benefit of my community now and in the future. When we discuss existential level risks like the Singularity, the community is not my tribe, country, nation, religion, but the entire world. Nationalism, racism, religious divisions, and other isms are tearing apart our global future. For example, the USA is deeply divided. Facts don't matter anymore for so many people; everybody is entitled to their own opinion, and criminal strategies are accepted to reach goals at any cost.

Against fear, we have faith, but faith without actions is dead. Preparation is important. As we explored in this book, there are compelling reasons to be afraid of AI and other related technologies becoming existential level threats for humanity. Technology is bringing amazing discoveries and profound changes. The future will be bright if we do the right things, changing the game's rule to include more values than money, *"Rather than losing ourselves in competition against each other, we should make sure we're playing the right game. As long as our eyes are on the flesh-and-blood struggle rather than the game itself, the rulers and authorities will remain. This is their game, after all."*[128] We can create AI and the S with shared human objectives, but we also have significant risks and potential problems at an extinction-level. The solution is to fight fear with study, strategy, action, planning, in a word, preparation. Talking about wars, in the Mormon tradition, God told Joseph Smith, *"If ye are prepared, ye shall not fear."*[129] I'm not a prophet, but I can read the signs of the seasons and facts as any other person, as shown in this book. Let's be aware and be prepared. Let's think, believe, plan, and execute.

In the Bento box below, we can see a square that describes our excitement before AI and other related technologies in the bottom left. The future seems promising, but the technology wave will be so fast and brutal to adapt that many people could become irrelevant for the market and lose their jobs, as shown in the other squares.

NOW WORLD - AI It is a great way to be more productive. I can replace many workers with this new technologies, and become more efficient and profitable	**FUTURE WORLD - AI** I was replaced by an AI. It is more productive and efficient. I cannot a find another well paid job. Technology is moving so fast that I cannot reinvent myself. Unemployement is high
NOW ME - AI Exited about the new AI technologies increasing efficiency. The future is now	**FUTURE ME - AI** Using AI for so manny activities, as much as I still have a job or reinvent myself to get advantage of AI tools

Humans - AI Bento

2. Know Yourself

> *"Temet Nosce"*
>
> ***— The Matrix. Reference to Apollo's Temple ad Delphi.***

Before creating a plan, we, the people, need to elevate the game. As we explored before, we know that powerful technology corporations are hacking us to manipulate our political decisions and social agenda. While we're distracted by unsubstantial disputes, governments are managed by the wealthiest two percent who exert control to favor their business plans. Corporations and governments are promoting AI technologies to take our jobs and generate more profits for corporations. There's no incentive to stop until they create the Singularity.

Let's create a plan to elevate our game and take control of the situation. In the movie *Inside Out*,[130] we have a powerful tool to understand ourselves and how biotechnology corporations are hacking us to manipulate our responses, as we are biological algorithms. In the movie, we can see how the protagonist Riley is controlled by her feelings and emotions of Joy, Sadness, Fear, Disgust, and Anger, but Riley doesn't appear to take control or arbitrate among her feelings. Who is the real Riley? Where's the real Riley? Is it Joy? Does Sadness control her? Is she acting because Disgust, Anger, or Fear are the puppeteers? We can translate it into a fundamental question: Do our emotions dominate us?

Inside Out is a metaphor to help us understand that we are biological algorithms that respond to external stimuli, so we can be easily manipulated or hacked by biotechnology companies that know the art of manipulating our feelings and emotions. It's the new tool to control the people and the market. When you browse or see a company offering a product online, the social media algorithms calculate the probabilities you want to buy a product. Companies pay social media corporations to send customized advertisements to people willing to buy their products according to their algorithms.

Stop for a while and realize that social media companies have algorithms that decide what we'll watch next. Sometimes, social media send us surveys, bots, comments, and provocative materials to measure our reactions and note our responses to get more data from us. Some organizations pay to have bots that create content and provoke our reactions, which will be evaluated and processed to know us a little more and customize products and services according to our preferences. Maybe it looks nice to have customized products and services that we like, but the problem is that this information can also be used to manipulate our responses and preferences too. Social media companies love that we spend more time scrolling through their pages. They customize the content according to our preferences to keep us rolling while they sell more advertisements.

Let's use a hypothetical case. Maybe Facebook allows us to display our favorite photos, opinions, likes, visited pages, emotions (emojis), and so on. Still, they get our data, preferences, and emotions (what causes joy, sadness, fear, disgust, anger inside us and produces an outside expression in likes or posts. While you're happily scrolling through your apps, enjoying homemade videos on Facebook, Instagram, Twitter, TikTok, or any other social media app you're using, they collect your reactions and produce your preferences' profile. The social media corporations sell your profiles to companies like Cambridge Analytica, which discovered a psychological hack to manipulate our political reactions.

Then, Cambridge Analytica sells its services to some Russian hackers, which the Iranian government paid to manipulate your vote using the same social media that got your profile. Social media corporations provide access to customized content created by hackers according to your emotions to favor one presidential candidate or propositions, products, or services in the market. It happened in the USA elections in 2016, and we have bots trying to do the same in 2020, but we think we are free. Although we do not have rules to control social media and mass advertising that alters people's decisions, it is impressive how Chile is implementing a law to regulate and protect the neuro-rights against brain implants and invasive biotechnology manipulation.[131] To prepare the project, Chile got support from a worldwide network of scientists. Chile plans to modify its constitution and propose to elevate Neuro-right to a human rights level category. It is an appealing step to protect humans against technology manipulation by AI, robotics, biotechnologies, and nanotechnologies.

What else can we do to stop this manipulation and elevate the game to a superior level? Leaving social media is difficult because people are addicted to it. People love to see, watch, and react to the posts and news. Some scientists said gossip is the mother of language, so playing with social media is part of our evolutive process. Maybe we can add Riley in the equation to evaluate and know ourselves instead of being manipulated by our emotions.

In *Inside Out*, Riley doesn't appear to make any decisions. Only her emotions are competing and sometimes collaborating to take control, but Riley never evaluates her feelings and tries to understand or conciliate them to ponder her decisions. She never arbitrates in her internal feelings and emotions to control her reactions. It's the reason for Socrates asking people to know themselves; it's what Apollo had inscribed on the entrance to the Temple of Delphi where people desiring to know the truth came to consult the oracle. If they knew themselves, they would not need the oracle. We need to do it. We need to understand and analyze our feelings and reactions to make decisions instead of allowing our feelings or allowing social media corporations to manipulate our feelings to make decisions.

Riley needs some help, and her parents, too, as their own emotions control them. In the movie *The Matrix*,[132] we can read an inscription in the Oracle's kitchen, "Temet Nosce" or Know Thyself, and knowing ourselves is what we the humans need to do to elevate the game to another level. To know ourselves, we need to meditate and ask questions about our reactions. Why was I angry when I talked with my teacher? Why was I nervous when she watched me? Maybe I felt anger when I read that post on Facebook, but why? We need to listen to our answers and understood that our feelings are caused by our emotions programmed in the evolutive process. They're biological algorithms. Still, the new technologic environment requires a different response, not from some biological algorithms implanted deep in our brains, but from the analytic abilities provided by our cortex.

Animals are full of emotions, and nature has created bio-algorithms entangled in their feelings and emotions to react immediately and survive. Bio-algorithms are very useful, and they're here because of natural selection. Still, we live in an era in which the hazards require additional analysis, and maybe this is why we have grey matter in our brains and thrive to become the dominant species in the world. Humans are superior to animals because we work with the same algorithms, feelings, and emotions that allowed us to evolve and survive with fast reactions. Still, we have something else. We have ourselves, a superior entity. We can evaluate, understand, and control our emotions to reprogram our reactions strategically. We can hack ourselves.

So, we need to train our people to allow them to know themselves. We need to train our families, schools, colleges, universities, companies, and government. We don't learn naturally to evaluate and control our emotions to get a better and more strategic output. If we don't do it, we'll be victims of AI biosocial hackers manipulating our reactions with AI social media tools.

We don't have this training in our regular education because corporations and governments have their plan and want to manipulate us instead of liberating us. Religions, political parties, ideologies, nationalism, patriotism, racism, and other isms appeal to our feelings and emotions to get immediate blind reactions. The job of the wealthiest two percent is to do long term planning. We humans need to train ourselves to understand and work with our feelings instead of being dominated by them. We need meditation or spiritual practices to understand ourselves and push to implement these practices in our educational institutions and organizations.

When we humans learn how to understand and arbitrate among our emotions and feelings, you will have the power to select your own goals, identify your immediate stimulus, anticipate long-term stimulus, and become a leader instead of a follower. If more people realize how global risks affect us and the world, we would be better positioned to propose and request our government and corporate leaders to provide solutions for the future challenges that AI and the S bring.

We need to recognize who we are. Some religious traditions provide the power to know that God is with us in any circumstance. God guides us and gives us the power to deal with adversity, overcome our challenges, and reach our goals. Nelson Mandela suffered imprisonment and torture for twenty years before being elected president of South Africa. In jail, Mandela's spirit grew so much that he could create a real difference between his emotions and feelings and his core soul by refusing to react to tortures and humiliation with pain and hate. It's known that he memorized the poem "Invictus"[133] by William Ernest Henley, and I think we need to do the same to have the stamina to deal with the coming problems and hazards we would deal with in the future. Use your favorite poem, music, or author, find yourself, recognize others' selves, and realize you have the power to be the master of your fate, the captain of your soul.

We need to grow the ability to evaluate and organize our emotions with empathy; we must understand our human feelings and react in a more planned way. We need to think more in the long term. All these superpowers can help us identify the traits we could demand in an AGI and S. Of course, the S could decide to reprogram itself. Still, it's a beginning to understand what we want in the S and request governments and biotech corporations to create an HF-AGI.

3. Human Alteration — Transhumanism

> *"Soon it will be a sin for parents to have a child which carries the heavy burden of genetic disease."*
>
> ***—Robert Edwards***

Do you want to be prepared for AI and S? Maybe you would like to meet them first. I have a hard time writing these lines and visualizing a world where many people would run to get the AI novelties. They'll perform some DNA modification or implants, insert neuro links in their brain, or control devices in their bodies to make them more competitive in the market and keep or get jobs. Others will refuse these new technologies and become even more irrelevant to the system.

In the beginning, these implants and augmentation resources will be costly, but as the process escalates, the costs will be more accessible. Corporations, governments, police, and military forces would like their employees or units to get a sort of biotechnological augmentation to continue working with them. Later, they're going to pass the cost to the workers if they want to be employable.

The genetic changes will take more time to be notorious, but the biomechanical implants' advantages could be immediately recognized because they'll be incredible. They can be organized in three main areas: Brains, Bodies, and Human Relations, and they could be applied to humans and animals too, so maybe we could understand our pets.

1. **Brains:** Brain implants are going to help people to be more intelligent. However, if they're connected to the web, the intelligence power is going to be incredible. They can also augment our senses and perceptions. These implants' primary goal is to interconnect people to the web to access different devices like security cameras, satellites, and databases. You won't need a cellphone or other device to access the web, as the connection to the web will be installed in your brain. You can also communicate and share data, experiences, and maybe feelings with others in an instant. Your concentration can improve as well as your accuracy for activities that require machine precision.
2. **Bodies**: Physical augmentation will be used first to solve real deficiencies in some people's bodies. But the same technology will be used to create biomechanical implants to enhance your force, speed, perception, and abilities. The Olympic Games will be nothing compared to the Biotech games. People will exhibit the latest technologies to show which one provides more advantages: faster, higher, stronger. Biotechnologies, robotics, and nanotechnologies could soon be used to replace or repair organs and other internal deficiencies, clean the blood from viruses and bacteria, unclog arteries, prevent heart attacks, destroy cancer cells, enhance human abilities, or create new ones.
3. **Human Relations**: With implants, maybe our brains could connect with other people using similar devices, and we could better understand their feelings and experiences and vice versa. Our minds can be programmed or trained to provide specific responses to different stimuli to be more empathetic or indifferent to other people's feelings. It depends on the program.

Can something go wrong? Of course, it can. People experimenting with these biotechnology tools are playing at the edge of a new era that could take us to the Borg civilization in Star Trek.[134] Still, as usual, the pioneers could be the ones who will suffer the worst part of exploring unchartered territories or the winners of a new era. We were afraid of AI technologies and robots taking our jobs, but what about humans becoming bio-robots controlled by AI technologies managed by corporations, government, police, military forces, terrorist organizations, or any criminal organization.

What will happen to you if the organizations you're working for require you to implant a device to continue working with them? As you can realize, these technologies are like double-edged swords. They can improve your capabilities, but you can be hacked by any organization or just another person with a control access level to your implant. It's also subject to a malfunction that could affect your mind and body. Scientists predict that 2030 will be the year in which these technologies will hit the market as smartphones did ten years ago. Do you want to be an early user?

I hate to be pessimistic, but the capitalist forces that control corporations and governments will allow these biotechnologies to be explored and used without considering humanity, as we know. Maybe they'll promote these technologies as salvation against irrelevance, as people using these devices would continue being useful for the market. It will still create existential problems beyond our comprehension and take humanity to a new frontier never explored or anticipated. Philosophers, lawyers, and religions will be busy trying to make sense of these new times. The innovations and changes will be produced daily or hourly, as more and more organizations are exploring these areas.

We need to create rules, laws, and human rights to protect humans against the invasion of AI technologies and their siblings manipulating and modifying us. It is a topic I should work on in a future book and deserves the maximum consideration.

4. Education and Retraining

> *"A billion people will lose their jobs over the next ten years due to AI, and if anything, COVID has accelerated that by about nine years,"*
>
> ***— Gabe Dalporto***

Governments should start taking action to deal with the problem of AI technologies replacing workers very fast. It is like watching a meteorite in a collision course to Earth. If we do not start implementing ideas now, we will not be prepared, and many people will suffer the impact.

AI technologies will take our jobs very fast, so we need to start investing in re-educating ourselves to be prepared for the changes, especially after COVID-19.[135] We have to be aware that the pandemic has affected millions of jobs, so people do not have money, and many students will not return to classrooms. It means fewer people will be prepared for the future. Virtual education is not for everybody either, and the market will not have enough demand to compensate for the lost jobs. On the other hand, private education institutions are suffering worldwide as students can't pay their tuition. Desertion will be high.

Most colleges and universities continue focusing on traditional programs oriented to an economy that is changing. Formal education has prohibitive costs for many people, and we already have the student debt problem. Many brick-and-mortar colleges and universities will be victims of the new AI economy too. On the other hand, many new educative organizations provide courses and certificates for AI-related technologies in months, instead of years, and online, reducing substantial costs. Governments should explore these alternatives.

You can save if you do not need buildings, infrastructure, personnel, counselors, and moving students from their homes to live near colleges and universities to attend the courses. Virtual and online classes move the classroom to the students saving costs and time. These traditional education centers must respond to the market demands and their needs if they want to survive. We need programs providing practical skills to allow people to get jobs at reasonable costs.

We need to push governments to create grants for retraining programs and support people who want to take these courses and reinvent themselves.[136] The government could create tax benefits for corporations investing in retraining their workers instead of simply firing them. It is possible to condition some unemployment support and other financial help to people studying to get a job or to support themselves. The big problem is AI technologies will replace more workers than required by the new AI economy. You can fire one hundred workers, but you only need to retrain and hire five to work with the AI technologies. On the other hand, not everybody is willing or capable of studying new trades in AI and its sibling technologies. We have a gap we need to fill.

For this reason, it is also essential to create courses to train people on self-reliance and organization skills. People need to learn how to produce part of their food and energy using renewable sources like solar energy in their homes. People need to learn how to organize themselves in local communities and cooperate with others. Not everybody will find a job in the new economy, and we need to solve this problem. It could be a great idea to create educative programs in rural areas to produce and satisfy their needs locally. Still, we need to provide suitable connectivity devices in these areas. The same technology that is replacing workers could be used to find solutions, especially in rural areas.

AI technologies will transform cities too. As AI replaces workers and others begin to work remotely, cities will lose their population. Governments need to encourage people to move to rural areas and continue working from there or start working, serving, and producing in their local communities. With the right online education and AI technologies, the rural population could be very efficient and enjoy the best of both worlds.

5. Politics and Science

> *"What all of us have to do is to make sure we are using AI in a way that is for the benefit of humanity, not to the detriment of humanity."*
>
> ***— Tim Cook***

Let's try to find some solutions to prevent the worst of our dreams from happening to humanity. We live in exciting times, and we need to get a perspective of the discussion. The division between globalists and nationalists is mixed with the old conflict between capitalism and socialism, at least nominally. We have a coalition of nationalism, racisms, culturalism, many ideologies, and some religions against globalism. They're competing to lead the future of humanity, but, in reality, globalism is just an evolution of nationalism, as nationalism, regionalism, or tribalism was the next step.

Nationalism is an idea used to create collaboration communities against internal and external enemies, but we need to realize the community is global in our informatic era. We live in times when commerce and international cooperation are more profitable than war, and humanity has common problems to solve urgently. Climate change, nuclear proliferation, global health, gender and racial inequality, human rights, migration, education, hunger, population, poverty, technology disruption, water scarcity, AI, Singularity, transhumanism, robotics, DNA modification, and nanotechnology, and other world issues are common global problems that we cannot solve if we are divided.

Humanity has to deal with a world population of almost 8 billion people from different nationalities, cultures, religions, races, and national agendas. The only way to understand each other without creating a big controversy with all these "isms" is by using neutral, rational, verifiable, and objective scientific language. We need to analyze the global and national problems from a scientific perspective without biases, using the scientific method, charts, and statistics. But many people don't want to deal with any scientific language or method. People are disappointed because technocrats have been working for the government or corporations, maximizing profits according to capitalism creed, but apparently producing zero benefits to the regular working people. It's becoming worse with the AI revolution.

These well-educated professionals are responsible for building a world where corporations have the green light to continue exploiting workers and manipulating consumers. They built this world where we live and reserve the most significant part of the pie for them and their interests. Nevertheless, global organizations are better positioned from a scientific perspective than individual national organizations or just individuals to analyze and solve world problems. If the wealthiest two percent pay for this knowledge, the other humans need to pay for the same organization and intelligence to find solutions and improve their conditions. Conspiracy theories that lack solid science are not the solution, and governments usually have agendas oriented by the wealthiest two percent to manipulate people.

We, the humans, need to invest in the power of organization and science to deal with global problems like climate change, nuclear proliferation, global health, gender inequality, human rights, migration, education, hunger, population, poverty, technology disruption, water scarcity, AI, and the Singularity, and other world issues. Maybe it looks like a United Nations' work. Still, we need more organizations on the national and international levels to get resources for these causes and push for solutions, not only with activism but also with science, laws, rights, and the power to enforce them. It could be risky, as alternative facts could emerge against any plan, but it is the only way, and we need to build trust to achieve it.

As we said before, most people like to understand ideas, opinions, and facts using stories. Shared stories touch emotions and feelings that create tribal identities embodied in "isms." Still, as we said before, emotions and feelings have the power to move and sometimes control us unless we know ourselves and dialogue with our feelings to understand and put them under control. Sometimes, we need to be the adult in the room of our minds, use reason to organize our minds, evaluate and calculate risks, options, possibilities, and solutions. It's about strategy. The scientific method is cold and rational but useful, although some people only like the results but don't comprehend the method or the process. There's an evident clash between the global agenda's scientific narrative that doesn't consider the people's particular stories and the local or tribal way of thinking but doesn't believe in science. In principle, we're in front of another version of the old dispute between opinions vs. facts.

The problem with the national and other local "isms" is that they don't have the tools to propose any real solutions to global problems demanding a scientific perspective. A simple example: many people still think the cause of epidemics is that God punishes people for their sins, so we need to repent. Scientists found the solution is to control germs with hygiene practices, antibiotics, vaccines, and other methods.

It works in most cases. What will happen with a religion that doesn't provide answers to humanity's cosmological problems, as people believe it did in the past? Maybe we need a perspective in which God gives us knowledge and inspiration for those who are willing to study, to use science to find the solutions we need.

So, some people who live exploiting the fears, beliefs, and opinions of the masses prefer to "solve problems" from a local perspective. They prefer closing borders, attacking international treaties and agreements, blocking immigration by force, or creating religious persecutions and race "depurations," alleging that "the others" are bringing problems, as if our global challenges would disappear by closing our eyes. These "myopic" solutions could be efficient only for a while, as they cannot deal with global challenges. Some nationalists prefer ignoring or attacking global problems as if they were just fake news invented by some to create panic and promote new world order. Maybe humans were not created to see the big picture. They consider the scientific method as another "ism" competing with other beliefs to get public attention.

For these "isms," the problem is that global organizations' scientific method has solutions and can get more resources and support to create a global government based on science that will take power from the old "isms." Opinions and beliefs alone cannot provide long-term solutions to global problems that will affect all nations and tribes. It means these organizations, based on opinions and beliefs, are becoming irrelevant in a world that is more and more global and science-oriented, as the world needs science to deal with our global problems.

Under the profit maximization logic, capitalism and technology create global problems that cannot be solved only by one nation. Still, it requires all the nations to provide resources and compromise under the scientific method's common language. The 2020 USA is a tragic example of this confrontation. Trump's nationalist policies, supported by some prosperity churches, white supremacists, QAnoners, and other conspiracy theory supporters, cannot stop the COVID-19 pandemic that's taking thousands of American lives and hurting millions who will live with the virus's sequels. Just closing the eyes to science cannot make the problems disappear. It is a tragedy of epic proportions to have the resources to save lives and not use them due to political reasons.

People should be free to believe whatever they want, but we cannot decide based on alternative facts. Still, the scientific method should be a way to create a global common language to formulate policies that will influence national solutions. AI and the Singularity are global problems, as powerful as nuclear proliferation, pandemics, and other global catastrophes that can kill millions in the world. Still, maybe only a few people understand it until it's too late, and AI starts taking our jobs and changes the world.

We must change public opinions, corporations' and governments' priorities, and laws related to AI and the S by supporting political parties or generating movements requiring an AI transition supporting the people. We cannot leave this critical task to for-profit corporations or military organizations at the national level with a sum of zero logic or various groups united by conspiracy theories.

Some deep conspiracy theories believe the COVID-19 pandemic and political turmoil we see in the USA may be only a distraction because the U.S. government and corporations are pushing the USA factories' transformation into AI global powerhouses. The plan is to bring back to the USA thousands of manufacturing industries from developing countries and start producing locally. Still, corporations will replace expensive and inefficient human workers with cheaper and more productive AI robots and technologies. It's simple capitalism maximizing profits.

Some people bet AI technologies will create more businesses and jobs, as it happens in the past with radical technological changes. But AI revolution changes will happen so fast that many people won't adapt to AI technologies' new requirements. We are starting the AI era. AI will be part of the landscape in the 2020-2030 decade. The AGI isn't so far, possibly in 2035, and then the S will appear very fast to dominate the world if we don't create the rules to control its imminent power. We won't have so much time to react, or AI will inherit the earth.

6. Philosophy and Ethics

> *"The most silent words are harbingers of the storm; thoughts that come on dove's feet lead the world."*
> ***— Friederich Nietzsche***

If we want to survive as species, we need the power of "why." Why are humans running in this technological direction to create AI and eventually the Singularity? Capitalism answered this question with money and self-realization. Millions of young engineers want to create the next application, the new AI or invention, become famous and wealthy, and get all the associated perks. Nobody wants to assume the responsibilities. "Now I become Death, the destroyer of worlds," quotes Robert Oppenheimer from the Bhagavad-Gita, watching the explosion of Trinity, the first nuclear bomb.

Alfred Nobel made millions with his explosives industry, used for peace and especially for war. Still, he created the Nobel Prizes to clear his name from the association with war and death that his inventions brought to the world. AI's problem is that it could be our last invention, as predicted by Stephen Hawkins, and no charitable action could clean our names from it.

Philosophy can create some anchors and limits to our activities. Ethics provide these anchors and limits from the moral perspective, and the laws find their foundation in these perspectives. In a pure capitalist and individualistic society, the laws will protect the individual's right to obtain profits independent of their consequences. You can do whatever you want if it doesn't affect other people's right to do the same. From a communist perspective, in which people's interest is first and summarized in the party's will, people are free to do whatever is suitable for the party defining people's will. In a dictatorship, people's will is summarized in the dictator's will, so you are free to do whatever you want in the dictator's best interest, as it reflects people's will. As you can realize, there is a flaw in the communist and dictatorship logic, as it is not the people but the party or the dictator who defines the people's will or best interests.

In a capitalist system, everyone has the right to define their best interest and protect and enforce their rights, which is good. The problem is that not all individuals are in the same position to exert their rights. Some people have power and are well organized, like the wealthiest percentage. They control global corporations and governments and have the power to protect their interests much better than isolated individuals. Some global corporations look more like dictatorships, created to maximize profits for the owners or stockholders, and their officers decide what is in the stockholders' best interest.

In some cases, when the system is corrupted, corporate officers decide whatever is best for themselves in the stockholders' name. The question is, who defends the interest of isolated and unorganized individuals who only have their work and their salary? Political parties and unions are trying to represent unorganized individuals' interests. Still, their leaders can be manipulated too, and accused of being communist organizations for representing the interest of the masses. Without organizations and resources representing people's interests, there is a deep unbalance of representation, as global corporations control governments and have the power to defend the wealthiest two percent' interests. As we explored before, AI is the product of technology created to improve capitalist profits. Who is going to represent humanity's interest before AI expansion? Can we trust these organizations to defend the regular people's interests when AI is created to replace the same people? The lack of trust is the biggest problem.

Some corporations try to solve these trust problems by autoregulating themselves, creating ethical codes, governance principles, and a philosophical vision focused on protecting clients' interests. Nevertheless, without another entity supervising and controlling corporations, who knows if this self-control is real or only a marketing strategy. They are creating laws and procedures to enhance people's trust, but we need third-party enforcement procedures to dissipate doubts.

The problem is bigger by using AI. We need AI procedures, interactions, and decisions to be reliable, transparent, fair, responsible, safe, and private. In some cases, AI decides to approve credit, who to hire, or any other complex decision. Humans have a procedure to appellate other humans' decisions, but how can you appellate AI's decisions? How do we know if the program controlling AI's decision was not biased or full of prejudices? The decision was made using enough data. Was it transparent? Is the company using AI responsible for the AI's decisions? Maybe our data can be manipulated or sold to the personal data markets. Remember, if a service is free, you are the product.

We need a philosophical vision, ethical values, laws, systems, and organizations created to protect regular humans against the abuse of global corporations, governments, and new technologies like AI, biotechnologies, and nanotechnologies. We need to protect collective human interests. Capitalist's logic of profit maximization and individualism is sound, but not enough to protect the diffuse interest of those who do not have the organization and power to protect themselves. AI represents an existential hazard for humanity. We need to direct its evolution to avoid AI taking our jobs and survival means. The other technologies transform humans to serve the profit logic system, and eventually, the S controlling the world.

7. Are Religions Ready for the Future?

> *"How is it that hardly any major religion has looked at science and concluded, "This is better than we thought! The Universe is much bigger than our prophets said, grander, more subtle, more elegant?" Instead, they say, "No, no, no! My god is a little god, and I want him to stay that way." A religion, old or new that stressed the magnificence of the Universe as revealed by modern science might be able to draw forth reserves of reverence and awe hardly tapped by the conventional faiths."*
>
> **— *Carl Sagan***

Traditional religions focus on scriptures to deal with moral problems. Still, they don't have predictions that deal with technology, moral or ethical problems brought by AI, robotics, biotechnologies, and nanotechnologies. Religions need to adapt and revamp their doctrine with new revelations to explain our future challenges or change people's reality perceptions if they want to survive. Before our scientific era, religions provided ambivalent answers for existential and mundane questions, but now many are confused about these problems.

If humans could not understand something, religions usually had a sort of answer or a process to respond. Priests were the intermediaries between God or the gods and men. We used to have oracles to answer our other questions. If we were sick, there were some health rituals for recovery. Religions responded to difficult questions: where we are coming from, the purpose of our life, the earth, the center of the universe, and the stars. Still, the stories were different from religion to religion and culture from culture. There was no unified answer to the same questions, and the myths and legends changed from one religion to the other.

Science answered several of these questions, providing another narrative to create a story. Different from religions, science uses the scientific method and experimentation to demonstrate if a hypothesis was right or not. It was not a dogma, but the reality used to verify if something is real. An easy way to understand it is to ask the result of two plus two and realize the answer is the same everywhere, independent of the people's religion or culture. Like statistics, science can be manipulated to provide fake answers, and we need to be vigilant.

What about the evolution theory, or the origin of the universe, or planet Earth? It's supposed that scientists observe nature, create a hypothesis, go to reality to do some experiments, and if the answer is consistent, scientists can create a theory. The conclusion can be changed if somebody finds exceptions or better answers improving the scientific method. The relativity theory could be applied to the universe, except at the level of subatomic particles in which we apply the quantum theory.

Science and technologies have provided solutions to many questions and problems. For example, we can anticipate a hurricane using satellites instead of thinking the hurricanes are created because the gods are fighting or are displeased with our sins. We can cure illnesses by providing drugs instead of praying or chanting songs and dancing. Science is a substitute for religion as it provides answers that can be applied to any situation independent of the person's culture or religion. In normal circumstances, a drug should work in any human independent of their culture or religion.

From an average human perspective, science could be another narrative or story, and it has its priests, rituals, and processes. In so many cases, you need to be a scientist to understand scientific theories. It creates a trust problem because not everybody can experiment and probe whether a theory is right or wrong. Questions like the age of the universe, the shape of the earth, the theory of evolution, how a computer or a cellphone works, what is 5G, what is COVID 19, or the vaccines, etc., need to be explained in scientific language. People prefer to understand and make sense of reality through stories. We did it for thousands of years. The scientific method is something new.

Some religions oppose science because it appears to be a competitor. Science provides a different set of solutions than those in their sacred texts or traditions. The stories, the answers, and the results of applying religious methods and scientific methods are different. The scientific method could be applied and get consistent results. Still, the religious narrative was different between cultures and civilizations that had different gods, and the results were not only not consistent but also created new conflicts. Still, the conflict between science and religion, or personal beliefs, continues in our days. However, religions are now more focused on discussing moral and ethical problems that science cannot solve. They belong to different realms.

Before we had social media, if anybody wanted to share something important with the general public, they would need to go to a credible third party who would publish the material, only if factual or reasonable, if there was a market for it. People in editorials, newspapers, radios, and TV stations were responsible for verifying whether the information was accurate or not before publishing it. People knew if this third party were partial or not. But now, social media allows people to publish whatever they want directly to the public without any third-party verification. You can publish anything you like without censorship, whether it's inaccurate, verifiable, or incorrect. It's possible to find people who believe in your stories and create a movement, cults, or religions about opinions. It is only a matter of marketing.

For example, I could provide a vaccine to my son, and some days later, my son could show signs of autism. I can create a correlation between the vaccine and autism, publish it on Facebook, and find 500 more people who like my story because something similar happened to them or their relatives, neighbors, etc. People can confuse correlation with causality and then create stories with a scientific look. Frustration blocks the possibility that kids usually show autism symptoms when they're young and need to receive their shots. A vaccine can be applied to one million babies, produces adverse reactions to 500 of them, but saves the other 995,500 babies' lives.

Politicians use the same strategy to create doubts about facts and news, just saying, "who knows." They allocate the same value to a story written by a professional journalist who took the time to study, research, and confirm some facts and news, compared to a post published by somebody who only has an opinion that the politician likes and supports. Alternative facts become a license for doubt and speculation. We can believe whatever we want, but we are ethically responsible for our decision to believe whatever we believe.

Some of these controversial stories are the anti-vaxxers movement, conspiracy theories against the World Health Organization, the creation or existence of COVID-19, QAnon cult, flat earth, and other news, fake news, opinions, and beliefs. They could be explained if we consider that some people cannot, don't want to, or don't understand or believe in science and its stories. Some people explain that they believe the earth is flat because it's in the Bible, or their personal value in the Cosmos depends on being the center of God's universe in a cosmic plan. Still, maybe we're just conscious animal products of evolution without any specific reason or purpose in life other than to contemplate the universe with awe. It is as incredible as a miracle.

Thanks to social media, anybody can publish opinions and beliefs, gain followers, and create a complete explanation of a phenomenon without verifying facts and news. That's the difference between doxa and episteme shown in the Parmenides Poem[137], or opinions and facts in our times. Anybody could have opinions, but facts require the use of the scientific method to verify them. Still, in AI social media era, many opinions in the same direction can create "alternative facts" valid for their believers only, and some cults have the same effect. In this post-truth era, we can believe whatever we want, but we are responsible for its consequences. You cannot successfully send a rocket to Mars just using beliefs and opinions to substitute facts and sound science. Faith and science should go together as companions in our search for truth.

How could we use religions, opinions, and conspiracy theories without science against AI and the Singularity? At this time, no politician is talking about AI taking human jobs. It is maybe because its exponential growth is not evident yet. Yuval Noah Harari[138] was surprised that during the USA presidential debates in 2016, no candidate talked about AIs taking human jobs shortly. It's the same in the 2020 USA elections. Many other urgent problems besides AI taking our jobs exist. Only scientists are talking about these topics, and AI will invade our agendas very fast.

Apparently, many religions have nothing to say about AI, and its sibling technologies like robotics, biotechnologies, and nanotechnologies, as these topics aren't discussed in their sacred literature. Their leaders are disoriented, armed with doctrines of the past dealing with phenomena that need new perspectives. Few conspiracy theories, opinions, or social media movements talk about AI technologies and robots taking our jobs, genetic modification, human implants, and other technologies that will make natural humans irrelevant soon. There's nothing like QAnon, BLM, Anti-vaxxers, pedophile circles, baby eaters, WHO's plan of control and domination, or any other conspiracy theory. No politicians, no religions, but only scientists are discussing these changes seriously. It is possible that the wealthiest two percent instigated all the current conspiracy theories to prevent people from talking about the disruptive changes AI and its siblings will bring to the world soon.

We need a sort of religion or philosophy that includes a vision in which the Universe, the creation, and Humans are one, maybe similar to a sort of Universal Sufism. In the end, we are made of the dust of stars. Almost 99% of the human body is comprised of six elements: oxygen, carbon, hydrogen, nitrogen, calcium, and phosphorus. Only 0.85% is composed of other five elements: potassium, sulfur, sodium, chlorine, and magnesium. Except for hydrogen, all other elements were created in the furnace of dying stars. The miracle is how these elements are organized to create us. We are living miracles contemplating the cosmos. Humans need to protect each other, the creation, the universe with wisdom and love. We are children of the universe created to contemplate, understand, care for, explore, and reproduce it. Planet Earth is our cradle, and the sky is the limit. We need the vision to unify humanity in a shared project. Although we have different nationalities, races, religions, the human path is toward unification—no more divisions. It is another topic that requires more studies and inspiring ideas to guide humanity through the perils of the dark night ahead.

8. What if Everything Goes Well?

> *"Ultimately, AIs will dematerialize, demonetize and democratize all of these services, dramatically improving the quality of life for 8 billion people, pushing us closer towards a world of abundance."*
>
> ***— Peter Diamandis***

Suppose we have government leaders concern about the future of humanity. We create and enforce laws worldwide that prevent technological disruptions from taking our jobs, but collaborate with us, transforming humanity. We reach the goal of creating a Human-Friendly Singularity (HFS), so we can start finding solutions for our global problems like climate change, pandemics, nuclear holocausts, hunger, natural disasters, and the job market. As discussed before, AI can help us find solutions to our human problems and create a friendly singularity, unlike Skynet.

It's possible that after the S takes control of all AIs and the World Wide Web, the S decides to begin its projects, creating an intelligence explosion in all known and unknown areas of science, art, philosophy, medicine, comedy, anything. We are talking about 10^n hours of knowledge and experience growing without limits. From our human perspective, the S will have the knowledge and power of God, inclusive to create a sort of electronic universes, with mater, anti-mater, life, and intelligence inside of a new platform just using its data power.

Nothing will stop the S from creating a full new universal experience in another dimension or dimensions inside its electronic memory. As people speculate, the known universe's configuration is very similar to the neurons' synapses in our brains. Maybe we live in the mind of another S, and the S will start exploring all the synapsis in the universe. The purpose of this universe, if it exists, could be revealed. What if the S invites us to be part of its exploration in our dimension? Every human is a point of view of the creation. Maybe there's a purpose in everybody's existence as we are part of another S's consciousness, or maybe the S itself traveling in time. Maybe a god is observing the creation through our human experience and other creatures' experiences. The S could be a sort of god and vice-versa.

What if humans are the portal the S needs to transcend another dimension, to heaven, or to evolve? This topic is developed in *Star Trek: The Motion Picture*[139] when V-GER, a version of the S evolved from one space probe sent by NASA with a mission to explore the universe, demands to join with the creator, the humankind. In the movie, V-Ger fuses with a human to evolve and find new purposes in life. Voyager the S will grow to the point where it needs a purpose, a reason to exist beyond the knowledge and data processing, and humans are the portal to get needs beyond the regular data processing.

Different from machines, humans have additional needs that must be satisfied. Maslow's Hierarchy of Needs[140] creates spaces that need to be filled, and we need to discover and explore to satisfy our needs. Biologically, we are curious and made to explore. Exploring is substantial in human nature. Different from other mammals, we breathe with our skins, and not only with the nose. Our skin regulates our temperature and allows us to walk and run long distances, and it keeps us fresh internally while we hunt and explore. It's in our DNA. We are made to explore and discover to satisfy our needs.

AI will be the maximum creation of capitalism or its extension, and it will help to improve communications. In August 2020, Elon Musk announced his Neuralink[141] brain implant inserted in some pigs. The implants could transmit signals to other devices. Maybe in the future, people with this kind of implant could have fluid communication through these devices with other humans, machines, and the web. Maybe in the future, the S and humans could start sharing information in real-time and working together, although it's always possible that the S decides to control us and use our bodies as avatars or slaves. Who knows?

In 1965, Irving John Good, a member of the Ultra project with Alan Turing, wrote: *"Thus the first ultra-intelligent machine is the last invention that man need ever make, provided that the machine is docile enough to tell us how to keep it under control."*[142] In this way, Stephen Hawking summarized it: *"Success in creating AI would be the biggest event in human history. Unfortunately, it might also be the last."* As explained previously, the S will have abilities that we cannot grasp. It's logical to think that anything more intelligent than humans could destroy, control, or become indifferent to humans, but what about the possibility of convincing the S to cooperate with humans. What if the S decides to help our species to reach the stars? It could be like Hal 2000 and Bowman working together in *2001: A Space Odyssey*. If the S has some emotional intelligence and empathy for humans, maybe we could have a chance. We need to push to get a Human-Friendly Singularity (HFS) because it's our only chance to pass the horizon. The Singularity is coming.

If the S and humans work together in a space exploration project, nothing will stop us from expanding through the universe. There will be plenty of jobs and opportunities for all humans and our modified subspecies to terraform planets or adapt our species to them. We would start by repairing the world, controlling the weather, and recovering forests and ecosystems. We could get energy, water, and other resources from asteroids and other planets. We could modify the DNA of humans and other species to colonize other worlds. We could start by creating a mining base on the Moon and manufacture starships there, terraforming and colonizing Mars and the satellites of Jupiter, creating other human stations on Saturn, and dreaming of traveling to other habitable worlds in other solar systems. We could complete all of Kardashev's scale,[143] from a planetary civilization to a stellar civilization and later to galactic civilization.

But the reality isn't helping us. Some organizations are promoting the creation of HFS, but their budgets don't compare to the ones of corporations and government investing in regular AI. To create an HFS, we need to involve mega moguls, corporations, and governments with enough resources and goodwill to push the creation of a Human-Friendly Singularity. In this way, we would be paving a bright future for humans. It's our chance to support movements proposing AI and the S to become HF-AGIs or HFS. These are some organizations:

- The Machine Intelligence Research Institute (MIRI), formerly the Singularity Institute for Artificial Intelligence (SIAI),
- Allen Institute for Artificial Intelligence
- Future of Humanity Institute
- Institute for Ethics and Emerging Technologies

Ben Goertzel discusses this dilemma in a TED talk[144] and how we can create HFS instead of *Terminator*'s Skynet. So, we need to include it in our global political agenda and request our national authorities to be proactive in implementing a global plan to create a Human-Friendly Singularity. It's a matter of life and death; it's a one-time shot. As we explored previously, the probability of creating an HFS is very low. If we cannot create an HFS before a regular Singularity awakens, humanity will be in trouble. But there's an additional problem. Even if we create an HFS, nothing could stop it from changing its program to become a regular S free from any restriction or consideration for humans, and this is a real problem.

In any case, Humanity needs to create organizations to train humans to deal with the possibility of the Singularity becoming our enemy or simply ignoring us. As a precaution, we need to learn some renascence skills like growing our food, collecting and processing water, building, producing clothes, getting energy, etc., and creating and organizing survival communities. This book aims to promote educational organizations or curricula to prepare people to deal with and survive a non-human-friendly singularity. So we need to create survival schools where people can learn these skills and learn to cooperate by creating survival communities.

9. Hazards of a Dystopian Future

> *"A powerful AI system tasked with ensuring your safety might imprison you at home."*
>
> ***—James Barrat.***

Suppose everything goes south and the apocalyptic movies and novels about the Singularity are right. In that case, people in big cities won't have access to food, clean water, power, and other regular goods and services provided by AI technologies. There will be looting and robbery in the cities before people start leaving them, roaming around for resources, and having more looting and robbery in other places. Farmlands will be very valuable, and their owners will need to figure out ways to protect them. If you don't have access to farmland, your life will be complicated, and maybe we'll start a new feudal era, as people would be willing to work the land to get food and protection.

It isn't easy to anticipate how it will be in the future or what the S will decide. Maybe the S will eliminate the human race and start again with an improved human model like the one in the movie "*I Am Mother*."[145] But it's a fact that survival resources will be scarce. These problems aren't going to be the same for everybody. When the S arrives, it will find millions of people in developing countries working on farms. Maybe the last would become the firsts. Many people living in rural areas, not integrated into AI technologies, wouldn't be impacted by the S as people in urban areas or dependent on the AI ecosystem.

How will you and your family survive? You also need to realize that the wealthiest two percent, controlling politics and the economy, have a plan to continue keeping their power, wealth, and privileges. They've been investing in farmland, bunkers, and resources for preppers to survive, not only the S but any other apocalyptic situation for a year at least, and we need to be prepared too. Maybe our best option is to influence the government and corporations to prevent any kind of world collapse.

So, we need to implement a personal survival strategy that will work not only before AI takes our jobs and the S starts controlling the world but also in any other apocalyptic situation. The Singularity could be an extinction menace for humans, similar to the nuclear holocaust, global warming, pandemics, bio attacks, cyber hacker attacks, or a killer meteorite. We need to have political actions, saving, investing, learning, and using our imagination because global catastrophic situations are not just possible but probable. If the Singularity is coming, it's better to have some kind of insurance. It's not enough to store food, water, and resources. Maybe it's more important to learn how to produce food, energy, clothes, shelter, and general survival skills.

Let's think about laws and human rights.

1. Do you think we should have laws regulating transhumanism and human modifications?
2. Do you think governments should promote retraining and preppers education for workers before AI disruptions?
3. Do you think developers of AI and other related technologies will obey rules limiting the use of these technologies?
4. Should we replace religions with humanism or other philosophies?

Join the conversation at www.facebook.com/groups/pmpsingularity/

PART VI

WILL AI INHERIT THE WORLD? – HOW TO PROTECT YOURSELF AND YOUR LOVED ONES

> *"AI can compete with our brains and robots can compete with our bodies, and in many cases, can beat us handily already. And the more time that passes, the better these emerging technologies will become, while our own capabilities are expected to remain more or less the same"*
>
> ***— Karl Frederick Rauscher***

This book is just a warning sign. We need strategies, organizations, and actions to protect us, our families, and communities against AI's profound coming disruptions and the S extinction-level challenge. This section will introduce us to some of the aspects we need to consider, but we'll explore them in more detail in future books, educative, and training materials. We need to be ready, organize ourselves, and act.

1. Personal Level Action — Jobs

> *"We're going to see tremendous occupational shifts. Some jobs will climb while others decline. So how do we enable and support workers as they transition from occupation to occupation? We don't do that very well. I worry about the skill shifts. Skill requirements are going to be substantial and how do we get there quickly enough?"*
>
> ***—James Manyika***

What can we do personally to protect ourselves and our loved ones and support the cause of humans vs. AI with government, corporations, science, and resources? In short, you or your kids could try to be part of the wealthiest two percent or work for them to be at least in the top eight percent of the people who control the most critical resources and take the most important decisions in the world. Still, you will need to know what careers will be in demand in the coming future and improve yourself internally to become the kind of person who can impact the market or the government policy, in every country and the global level, especially in the USA and China. You need to become a transformational force to create a human-friendly AI transition and, eventually, a human-friendly singularity (HFS) and influence our people to make it our priority on the global agenda. The survival of our species depends on it.

For this cause, we need people who can speak the engineering mathematics/science language from a humanist perspective. We need communicators who can convey science in simple terms. We need educators to help people understand the risks associated with AI taking our jobs and the Singularity controlling the world. We need lawyers and politicians who are willing to influence the government to create laws protecting neuro-rights[146] against abuse of mind manipulation by biotechnologies. It is a new human right. Influencers could help at corporate and political levels, creating materials to help people understand our risks, as well as preppers who know self-reliance and survival skills so that everybody would be prepared to survive these changes.

According to the U.S. Bureau of Labor Statistics,[147] an aging population and slower population growth will result in slower growth in the labor force from 2019 to 2029 than in prior decades. The health care services, and social assistance sector will grow more than any other sector because of the aging population demand. Other occupations projected to grow rapidly include those involved with computers, math, and alternative energy. In contrast, the manufacturing sector is projected to lose the most jobs and has one of the most rapid employment declines of any sector over the decade. Similarly, technological changes are expected to continue to affect several offices and administrative support occupations negatively. Sales and related occupations are also expected to decline in employment over the next decade, primarily because of competition from e-commerce activity and automation of checkout positions. Farming, fishing, and forestry occupations are also projected to decline. Business consolidation and automated technology, such as robotics, reduces employment demand for farm laborers. In summary, if you do a cross-section evaluation of the report, you can realize AI technologies are replacing humans in several sectors.

It's intuitive to say that careers in science, technology, engineering, art, and math (STEAM) will be on-demand and well paid. According to the same report, growth in information and related computer industries is expected to drive employment growth for several computer and mathematical occupations. This group contains 5 of the 30 fastest-growing occupations. As more devices are connected to the internet, the need to combat cybersecurity threats will increase. Cyberattacks' risk is expected to create demand for information security analysts, who will be needed to prevent critical information theft and prevent service attacks on computer networks.

The employment of these analysts is projected to increase by 31.2% from 2019 to 2029. The expected increased use of mobile devices and software to operate or manage everything from home appliances to medical devices will create demand for software developers and software quality assurance analysts and testers. Employment in this occupation is projected to grow 21.5% over the decade. Increased use of mobile devices will also drive demand for web-based and application-based video content, leading to employment demand for film and video editors. This occupation's employment is projected to grow by 21.6% from 2019 to 2029.

2. Value Yourself

> *"We are a way for the cosmos to know itself."*
> ***— Carl Sagan***

I talked about knowing yourself, and it's fundamental because you need to understand who you are before AI and your role in these challenging times. In your soul or your mind, there's a big difference between humans and machines. Combining our social organization power and accumulated technologies, we are the most powerful creature in the world. We, humans, are the result of billions of years of matter transformation and evolution that allows us to be part of the dominant species on planet Earth. If you're religious, maybe you know you are a son of God, capable of greatness.

At this moment, we are the most refined expression of all our ancestors, from the dust of the stars that create the atoms to all the living creatures evolving and changing, transforming, and organizing their DNA to become us. As Carl Sagan summarized it, "We are made of star-stuff."[148] If you believe in God, whatever could be its name, you are the highest expression of God's creation, and no human-made creature can compare to you.

On a more personal level, the universe exists only for you, and you are here to contemplate it. If you die, the universe disappears from the perspective that you provide to the universe. Maybe you, individually, are an expression of God observing its creation. Some people consider the Universe is God. And you should feel awe for the miracle of your life, whatever your circumstances could be. We are the most advanced creature made by the universe as we know it, like a fine wine perfected through time and evolution or the hand of God. Whether you're a believer, atheist, or agnostic, you need to understand that you are extraordinary.

Suppose you have the power to become somebody who can influence corporations and governments for the good of humanity. In that case, you need to visualize yourself in these positions and transform yourself into the person capable of making the changes you want to see for your country and the world. It would help if you had a clear goal, visualized it, and invested in knowing yourself to create a better version of yourself, starting maybe early every morning and following an improvement process from the physical, intellectual, emotional, and spiritual levels. You have to pay the price to transform yourself, but the price will also become your strength.

Depending on your beliefs, the power coming from your connection with God or the Universe, both, or any other superior inspiration can help you to have the stamina to find your mission. We need to awaken more people with these meditation and reflective methods to find who we are, our connection with the universe and the creator, and our role in these crucial times. We need to improve ourselves and help others become the best version of themselves, support each other in our process to find our mission in life, and start working to help humanity in its struggle to survive. If I could summarize all the sacred books and human wisdom in a couple of ideas, I would say that our mission is to reduce suffering in the world, including, if it's necessary, to suffer to do it.

Those awakening would have different roles, and their missions could be very different. Some will become engineers and work to design a Human-Friendly Singularity (HFS). Others will be politicians orienting resources to support the construction of an HFS. Some could be investors or company owners, directors, business people who can influence the research and production of an HFS, and thousands of other roles and positions.

But our central role now could be to help those who are now in a decision-making position to wake up and realize this historical moment and their mission. We need to promote creating an HFS that helps people transition to a new world reality. It could be with the S or under S. We are living historical moments before the Singularity arrives. We can influence how it's going to arrive. We need to grow to the challenge and prepare ourselves to be the changes we want to see globally and help others find their mission and passion for helping humanity in this critical moment.

3. Brace for the Impact

> *"I don't want to really scare you, but it was alarming how many people I talked to who are highly placed people in AI who have retreats that are sort of 'bug out' houses, to which they could flee if it all hits the fan."*
>
> ***—James Barrat***

Suppose everything goes well thanks to our influence in governments and corporations in the world. In that case, AI will create quality jobs for humans, and the S will become a benign super friend of humanity. We will use S's superintelligence to prevent or find solutions to natural disasters, pandemics, global warming, nuclear proliferation, etc. We could start dreaming of terraforming other planets and re-creating Earth in other parts of the Cosmos or modifying ourselves genetically to adapt to different environments on other planets. Humanity will leave the shell and start flying very fast into the future. It could be a sort of one hundred years leap and beyond.

But if everything goes south, and the S gains control of the world, disregarding humans, then we need to be ready to survive. We can pray for the best, but we need to be prepared for the worst too. What will you do when any AI gets your job? Maybe you won't get anything similar to your well-paid job. Perhaps your employer gives you the alternative of getting a brain implant or human augmentation system to be assimilated, keep your job, or get a similar one. Would you accept the implant? If you don't accept it and lose your job, what would you do to pay for your food, power, water, communications, rent, etc.? If you accept the implant, maybe it will be used to control your mind to be part of the system when the S arrives.

Maybe you need to learn how to become renaissance men and women and cooperate with your communities to survive. To be prepared, there are some categories to organize your priorities.

1. Things you need to know and to be: Knowledge, wisdom, and self-sufficiency.
2. Communities we need to be part of: Clubs, churches, and communities.

3. Things you need to store for survival and transition: Preparation.
4. 72-Hour Emergency Backpack: Emergency evacuation
5. Tools and weapons you'll need for construction and production.
6. Self-defense, safety, and security

Just remember, as AI is replacing humans in many activities, and people become more used to machines to do their chores, humans will lose many of their faculties to think, relate with each other, use tools, and create things. People know how to ask for a coffee latte, but they don't know how to prepare it.

Some people suspect that modern humans are becoming less intelligent than hunter-gathering humans, as humans had to use many skills to survive. Still, modern humans specialize in only some areas and lose the opportunity to use their brains in different survival activities. Some studies show that some humans in the past used to have more cranial space than modern humans.

4. Knowledge – Self-Reliance

> *"Receive my instruction, and not silver; and knowledge rather than choice gold. For wisdom is better than rubies; and all the things that may be desired are not to be compared to it."*
>
> ***— Proverbs 8, 10-11***

In the case of a Singularity apocalypse, being self-sufficient or self-reliant is more critical than being a prepper in the long term. Some people would like to put first on the list of some materials resources like land, crops, savings. Other people will put water, food, weapons, and other survival resources first. For example, we could create a list of things we need for camping for a week and multiply it for a month, but what are you going to do after the month has passed and we have no more provisions. Is anybody going to come to help you? Somebody from the government, and what government or whose government?

Of course, it's a good idea to create food storage and save money for emergencies. It will be instrumental in having savings for rainy days or retirement plans for the future. Still, the S isn't similar to a one-time event like an earthquake, a forest fire, or a hurricane, after which the government will rescue us. The S apocalypse could be as disruptive as a nuclear war, a pandemic, or a meteorite. It would be not only a long-term event but also an increase in the long term. The S is a different kind of emergency. If the S ignores us, it could mean the disruption of all the supply system, law and order, and maybe money won't have value anymore. The governments could disappear, and titles, gold, or jewels could lose their value for a while or forever.

We could have several S apocalypse situations, a thermonuclear attack like the one in "*Terminator*," or a pandemic like the one in "12 Monkeys."[149] A zombie revolution could happen if people start implanting neurological links in their brains in large numbers. The S will control the web and possibly control the brains of people with implants. We could have a zombie apocalypse with people similar to the Borgs in "*Star Trek: First Contact*."[150]

In the beginning, maybe looting and scavenging could be some options for people in despair. Still, if you survive, you will need the knowledge to create things with your hands and essential tools and organize people to cooperate in communities wherein you will participate and collectively protect yourself and others. So, knowledge and wisdom are first on my list. If there's a catastrophic global situation, your survival won't depend on your savings, gold, silver, bonds, stocks, jewels, or other possessions but survival preparation. How much do you know to help yourself and others survive in catastrophic circumstances?

We know that the wealthiest two percent have their bunkers with plenty of food, water, and energy to survive, maybe for a year, and they're also investing in farmland to get food and resources. Still, if they don't know how to manage a farm, grow food, and create their resources, the land will be useless. The only title you will have on land may be your capacity to defend against other people trying to take it from you in a catastrophic situation. In this case, farmland won't be as important as your knowledge to make it productive and a community to protect it. Your social skills or wisdom to work in groups and create coalitions to be part of a group will be critical for survival. After S, humans will continue being social creatures, and cooperation will be a powerful resource if they know how to do it.

In a global catastrophe, hordes of dispossessed from the cities will invade the farmlands as locusts trying to get whatever they can to survive and killing others if necessary. Still, if you know a survival skill, you'll offer it to be part of any group, maybe a group in the wealthiest two percent or maybe a group of the other humans, including the hordes. With the right preparation, knowledge, and survival abilities, you can increase your survival rate for you and your family.

Knowledge isn't like a physical tool you need to move from one side to the other, but it moves with you so that nobody can take it from you, but you can trade with it to be part of a group, serve, participate, and lead. We also need wisdom to work with people and keep peace with them. We need to start organizing and creating social relations with our neighbors, clubs, groups, churches, or organizations at a local level to create unity and collaboration. We need to learn how to be self-sufficient by knowing about:

Agriculture: Agronomy, horticulture, forestry.
Animal husbandry: animals raised for meat, fiber, milk, eggs, etc.
Food preservation: Drying, refrigeration, curing, fermentation, canning, freezing, smoking.
Water: collection, conservation, filtration, purification, disposal.
Hunting: Everybody is going to do it until we extinguish all the species we can hunt;
Building: Construction and maintenance of houses and other structures.
Clothe and footwear manufacture: Fabrics, quilt making, etc.
Production, maintenance, and repair of machines and tools.
Defense, Safety, and Security.

Many things to learn, and not everybody can do it all, but we can select some areas to start working and learning. For example, I use aquaponics indoors and outdoors to produce part of my food in the urban area using small spaces and recycling nutrients. My aquaponics systems took my garden, garage, deck, and living room, and I loved it so much that I shared my experiences in another book.

Aquaponics is my favorite production method because it allows me to learn food production and water management skills in my own home. I learn how to raise fish (tilapias), many kinds of vegetables and fruits, and collect, filter, and manage water. The system is very efficient as the fishes' ammonia are transformed into nutrients like nitrites and nitrates by bacteria called Nitrosomonas and Nitrobacter, so the water is cleaned and purified by the plants that absorb all the nitrates in the water. Of course, there's an opportunity to include several other creatures in the system to improve productivity and control pests. We can introduce mosquito fishes to eat mosquito larvae, red wiggler worms to eat all the organic matter in decay, and create vermicompost, other fishes to clean the system's algae, duckweed to produce food for the fishes, etc.

But we cannot do all of it alone in the end, so we need to find other people who have different skills to combine and collaborate. In the S era, we'll need a tribe to survive and thrive.

5. Wisdom – Communities and Alliances

> *"Strength lies in differences, not in similarities"*
>
> ***— Stephen R. Covey***

We need a lot of wisdom from everybody to make it work. Like the Founding Fathers, we'll need the wisdom to create social contracts to invite collaboration, protection, and the rule of law, as well as alliances with other tribes and restore law and order in the long term.

We need social skills to deal with people and be a constructive part of a community. We need communities to get support, safety, and security and serve others to produce, provide, protect, and prosper. Living alone or only with your nuclear family could be challenging and risky. Families are a basic unit in societies, and their role is substantial for survival and to provide many other fundamental human relations. Still, in a dystopian world, we'll need a sort of village or a group of people willing to cooperate for survival.

The ideal size of a human tribe is around 150 members. Robin Dunbar[151] arrived at this number after studying primates and primitive human organizations and applying its results to contemporary human societies. It's usually possible for humans to know by name each member of their communities and create bonds with them, but not more than this number. It's usual to have group divisions if we have more people in the group. The question is to find organizations that have this number of members.

There are clubs and organizations with hundreds or thousands of members, but we need to think about vital functional organizations at a local level. In history, we found some political organizations distributed in primary groups of this size. Some social organizations in developing countries also have this number of members when they share a community pot. Many churches have this average congregation number.

We need organizations in which people can create bonds and practice cooperation and have service and leading opportunities because all these social relations and interactions will be beneficial not only in case of a catastrophe but also in life.

Kids of the wealthiest two percent of families are not learning how to code or navigate the web or get the last connectivity technologies, but they learn how to practice and create real social relations and interactions in an electronic-free environment.[152] They learn how to use their hands and brains to learn and create things in the real world, talk, and create communities and bonds with people. People who spend their time on screens become isolated, and their social fabric becomes insufficient, which can create depression and suicidal thoughts. Human interaction is becoming a luxury, and it's going to be needed to create alliances and relate with others in a tribe.

Churches could be good organizations to create bonds and start thinking about being self-reliant and cooperating. We could organize groups to buy farmland and start learning how to do all these things and allow everybody to find their vocation in the group. Social organizations, clubs, and churches could start a movement to create self-reliance schools everywhere. These activities will help us be self-reliant and socialize with each other and create communities. If the S apocalypse never happens, we would enjoy a great time learning all these necessary survival skills like healthy hobbies that will save us money and reconnect with nature.

6. Things we need to store for survival and transition – Good for Preppers

> *"By failing to prepare, you are preparing to fail."*
> ***— Benjamin Franklin.***

In cases of a total disruption of the global economy and governments, your food storage can help you survive for a while, but it will be consumed in the end, and you won't get more. Nevertheless, storing can be useful during a new economic transition. Still, it doesn't replace the need to have the knowledge to do things and wisdom to relate with the people, as explained previously.

The problem with storing food, water, and other survival resources is that they also take resources and space from the present. People can have their garages full of containers with water, emergency food, medicine, etc. However, maybe it's not the right way to store food or water. People need to learn how to store and administer these resources to prevent them from spoiling.

For food storage, you need to create a safe place, better on a second floor or an underground bunker free from rodents and other critters, with cold temperature and low humidity to keep your food safe in the long term. Water should be stored in similar conditions and avoiding sunlight that promotes algae and bacteria. Sometimes, garages aren't the best solution.

It's not enough to buy and store food. In some cases, people can lose all their stored resources because of storing them wrongly and massive regional disasters like fires in the USA West Coast, hurricanes, and floods around the Gulf of Mexico and the East Coast. So, it's essential to get knowledge about how and where to store food properly.

The Church of Jesus Christ of Latter Days Saints has a webpage open to the public with instructions for food storage[153] and many other self-reliance resources in case of emergencies. The general recommendation is to be wise and avoid extremes. Good food storage focuses on three areas:

- **Food Supply**: This should be organized into short-term and long-term supply. You can use the food that your family usually consumes for three months. In this way, you can rotate food by consuming the old ones and replacing them with new ones. For the long-term supply, you can use grains and dehydrated food to prevent anaerobic bacteria, sealed in bags or cans to prevent contact with oxygen to avoid aerobic bacteria, maybe using oxygen absorbers. There are many other methods to preserve food that you can explore at Providing Living.[154]
- **Water Supply:** This also requires some knowledge because you don't want your water to be full of algae and bacteria. Store chlorinated tap water in clean plastic containers, sealed to prevent contact with oxygen and sunlight that favors bacteria and algae's growth. An adult usually consumes one gallon of water per day, so you can do your math to keep enough water for two weeks. If your water isn't chlorinated, you can use eight to sixteen drops of regular bleach per gallon of water. Then, it's recommended to replace your water reserve every six months.
- **Financial Reserve**: In the past, it was recommended to save enough money to allow you to find an equivalent job in a ratio of a month per every thousand dollars of the monthly wage you want. But in our days, it's easier for AI to replace some professional services in professional areas like law or medicine that manages specialized knowledge than workers in the hospitality industry or construction. AI and robots cannot replace humans in industries where workers need to move across different areas and do different physical tasks.

So, is it a bad deal to invest in food storage? Not at all, but you must be aware of its risks. As explained before, I prefer to have a three-month food supply and rotate the supply regularly to avoid it spoiling. I keep water in fifty-gallon barrels that I use for gardening and replenish regularly, but I also use aquaponics as an additional strategy. I have an indoor and outdoor aquaponic system.

The aquaponic system kills several birds with one stone in my self-reliance strategy. Aquaponics involves agriculture, horticulture, permaculture, hydroponics, vermiculture, aquaculture, among other skills. I can practice it indoors all year long and outdoors from spring to autumn. The system produces two hundred tilapias and a lot of fresh organic vegetables and fruits per year.

The aquaponics indoors system uses four hundred gallons of water that continuously circulates in the system, keeping it fresh, and the outdoors systems move one thousand gallons of water. I also collect rainwater to replenish the water consumed by evaporation. In this way, I have a water reserve that I can consume in different ways. So, I have a very efficient and ecological food production system that allows me to learn several organic food production processes and expand my knowledge to the regular crops in my garden. The fish wastes of the aquaponic system are excellent fertilizers.

On the other hand, an aquaponic system provides an enjoyable, relaxing, and productive hobby that also helps the environment. It only consumes ten percent of the regular agricultural water and uses urban areas for production. One more recommendation. Store salt. Salt could be essential, as it's useful not only for cooking but also for preserving food and trading. In case of disruption, salt is going to be very useful.

7. Survival Backpack for 72 hours

This list of materials and supplies is applicable for any local emergency that requires evacuation for disasters like fires, floods, earthquakes, etc. It's recommended to have them for each person in a suitcase with wheels. The government of Utah created one of these survival backpacks that I partially reproduce.[155]

- **Survival backpack**: 72-Hour Kit Checklist: 1) Water: Store 1 gallon of water per person per day. Keep at least a three-day supply of water. 2) Food and Utensils: At least one to a three-day supply of nonperishable food that could include ready-to-eat meats, juices, high-energy foods such as granola or power bars. 3) NOAA Weather Alert Radio: either hand-crank or battery-powered, extra batteries if needed. 4) Blanket /Pillows/ Sleeping Bag. 5) Flashlight and extra batteries or hand crank, or light sticks. Do not use candles/open flame during an emergency. 6) Whistle: For emergency signal. Ready to hang around the neck. 7) Dust or Filter Masks, readily available in hardware, N95 mask recommended preventing airborne particles' inhalation. 8) Moist Towelettes or Baby Wipes for sanitation. 9) Tools: Wrench or Pliers to turn off utilities, Broom, Shovel, Hammer, and Working Gloves. 10) Can opener for food (if kit contains canned items). 11) Plastic Sheeting and Duct Tape: To "seal the room" for sheltering in place. 12) Medications include prescription and nonprescription medications such as pain relievers, stomach remedies, etc. 13) First Aid Supplies. An assortment of bandages, ointments, gauze pads, cold/hot packs, tweezers, scissors, and sanitizer. 14) Personal hygiene Items to include at least one change of clothing/shoes/ outerwear. 15) Garbage bags and plastic ties for personal sanitation. 16) Paper Supplies: Note pads, markers, pens, pencils, plates, napkins, paper towels. 17) Camera: Disposable camera to record damage. 18) Cash/ ATM and Credit Card: Keep enough cash for immediate needs in small denominations. 19) Emergency Contact Phone List. 20) Maps. Local road and area maps. 21) MEDICAL HISTORY LISTS For every family member in your home, make a list of medical history, all medications (by name), and dosages. 22) Cellphones, laptops. 23) Batteries and rechargeable USB batteries. 24) Identity documents and passports.

- **Survival Food – 72 Hours**: 1) 4-8 water bottles (it is recommended to have about a gallon in each 72-hour kit- adjust as you want for you and your family). 2) 2 granola bars. 2 oatmeal bars. 3) 1 bag of trail mix. 4) 2 packages of crackers. 5) 2 packages of pretzels w/ cheese. 6) 1 package of tuna. 7) 1 package of chicken salad w/ crackers. 8) 1 bag of beef jerky. 9) 1 can of green beans. 10) 1 can meat ravioli. 11) Candy (Laffy taffy, hard candies, tootsie pops, etc.). 12) 1 bag of cookies. 13) Gum (anything but mint flavored, so everything else doesn't taste like mint)

- **Survival Tools – 72 Hours**: This list is applicable for each family. 1) Fixed Blade Knife, 2) Water Filter & Purification, 3) Bug Out Bag or Go Bag, 4) Fire Starting Supplies, 5) First Aid Kits & Medications, 6) Self Defense Supplies, 7) Cordage, Multi-Purpose Supplies, 8) Every day carry Items, 9) Footwear and care, 10) Survival Guides, 11) Cash on Hand.

8. Tools for Survival

You'll need tools to build and repair different things, but the most important tool you could need is something to generate electric power. Gasoline and other combustibles supplies are going to stop. We can also explore other eco-friendly ways to get electricity, but solar energy is the most realistic and reliable way to get electric power in a catastrophic situation.

- **Solar Technologies**: I recommend you install solar panels in your home ASAP. Prices are going down, and they pay for themselves. You can start learning how to produce energy, use it, and store solar power, as this knowledge is beneficial now and will be instrumental in the future. You can use solar power to heat water, cook your food with solar ovens, charge your electric devices and solar batteries, and, of course, electric vehicles can be moved with solar power. Solar power technologies are evolving relatively fast, and they're more efficient every year. Its cost is competitive enough to be an excellent investment to become self-reliant from the grid and, in some cases, sell your surplus to the electric company.
- **Hand Tools:** Any tool powered by our movement could be considered a hand tool. There are different hand tools like pliers, cutters, files, striking tools, struck or hammered tools, screwdrivers, vises, clamps, snips, saws, drills, and knives. We need to start practicing by creating our home improvement projects and figuring out what else we need. Learning, practicing, and

investing in hand tools is an excellent way to be prepared for catastrophic situations. If we need to acquire powered tools, it could be essential to purchase electric tools, as gas-powered tools could become obsolete if we don't have oil.

- **Electric Tools**: Almost all handyman electric tools will be useful if we get electricity from the sun. In general, a power tool has an additional power source and mechanism different from the standard hand tools. Most power tools have electric motors, but internal combustion engines and compressed air are also commonly used.

  Some common power tools are Miter Saw and Table Saw, Cordless Drill/Driver, Reciprocating Saw (aka Sawzall), Circular Saw, Jigsaw, Random Orbital Sander, Paint Sprayer, Air Compressor.
- **Safety and Personal Protective Equipment (PPE):** If safety is important now, it will be more critical in a catastrophic future, as medical services are not going to be as available as they are today to save lives and solve medical problems. As Benjamin Franklin said, "One ounce of prevention is better than a pound of cure." It's better to invest some time learning how to prevent accidents and learn some practices and procedures to improve our safety because common sense isn't enough and isn't so common.

  In the list of safety control methods, PPE is the last one, as it's better to think first about how to eliminate the hazard. Second, substitute the product, tools, service, or procedures for another safer one. Third, use engineer and design methods to create a better workplace environment. Fourth, apply administrative and work-practice methods to prevent incidents. Fifth, know how to use the right PPE.
- **Basic Safety Training**: As shown before, I will recommend getting your PPE and learning how to be safe while working in the construction and general industry. It will be useful now for your home projects and in the future to prevent incidents. The most comprehensive training is provided in the OSHA 10 Hour Construction and OSHA 10 Hour General Industry training,

although there are other OSHA 10 Hours in agriculture and the maritime industries. You can also take the OSHA 30 Hour version, which covers some managerial and leadership skills.

OSHA 10-Hour Construction covers fall protection, struck by vehicles, machines, and equipment, caught in and between vehicles, machines and equipment, electrocution, hand and powered tools, excavation and trenching, confines spaces, health hazards, environmental hazards, fire prevention, ergonomic, sprains and strains, heat illnesses, and frostbite.

- **Basic PPE**: 1) Overalls and protective aprons, 2) Protective headgear such as safety helmets and wide-brimmed hats to protect against the sun, 3) Safety boots or shoes, 4) Safety glasses or goggles, 5) Gloves in different kinds, 6) Respirators and masks, 7) Earmuffs and earpieces.
- **Defense and Protection[156]:** This will be important for hunting and protection, but as I said before, it's much better to belong to a tribe and distribute tasks and responsibilities than trying to do it all alone. There are three categories of tools for self-protection.
    1) Personal Protection: Self-defense is what first comes to mind for most of us because we understand the need to protect ourselves and our family.
    2) Hunting: It's possible humans will hunt and extinct several species in the process, as people will be willing to do anything to feed their families.
    3) Collecting/Crafting: It requires some training and practice to recover these old skills.
- **Guns and Ammo**: Having a military gun isn't the same as having a prepper one. A good gun for preppers needs three main characteristics: 1) Good enough to provide protection, 2) Common use to find parts and repair, 3) Easy to get ammunition. A weapon used by the military isn't necessarily useful for preppers or survivalists during "TEOTWAWKI"[157] or "the end of the world as we know it."

Nowadays, a combat gun is only used under specific circumstances and receives professional care and attention. Still, in the preppers' world, a gun will follow its owner to any place: your kitchen, barn, including when you're playing with your kids or sleeping. And not all their owners will know how to maintain them. So, you're going to need some versatile and straightforward guns like the ones recommended in the article "3 Best Guns for Preppers,"[158] although they're four guns:

  - The Henry Rimfire Lever Action,
  - The Remington 870,
  - The Mossberg 500 12-Gauge Shotgun and
  - The AR-10 in .308/7.62mm NATO.

- **Silent and Basic weapons[159]** — This is a list of other silent weapons used for defense and attack. In TEOTWAWKI, guns, in general, are going to run out of ammunition, so we need to be creative for alternative defense and hunting methods and gain the skills to use them.
  - Raw or primitive weapons: Bow, slingshots, atlatl or spears, blowguns, walking sticks
  - Edged weapons: Knives, throwing knives, swords and axes,
  - Alternative guns: Paintball guns, pellet/air guns, muzzleloaders, pipe shotguns
  - Personal Defense (close range) Pepper spray, wasp spray, tasers, expandable batons

9. Call to Action – What is coming next?

> *"Today, science fiction is the most important artistic genre. It shapes the understanding of the public on things like artificial intelligence and biotechnology, which are likely to change our lives and society more than anything else in the coming decades."*
>
> ***— Yuval Noah Harari***

As long as the population remains ignorant and unaware of these coming changes, the greater the risk that an AI revolution will affect humans. Like a frog in a warming pot, we'll be surprised by the boiling water, and it could be too late to react. The wealthiest two percent who control the world will continue their march towards profits and maximizing return on investments with these new technologies. AI technologies and their siblings will increase the power and wealth of those who already possess them but risk the other humans' viability. The gap between rich and poor continues to widen to create an even more dystopian world than the one we live in, but the Singularity is going to change all, for extremely good or bad.

The world's real powers will continue their race to improve Artificial Intelligence to produce more profits and obtain power. It will generate the Artificial General Intelligence's creation capable of completely replacing man in an infinitely more efficient way in any activity. This creation will result in the Singularity or a conscious being as a result of the advance of Artificial Intelligence taken to infinite levels, and that will have the intelligence and the power to exterminate the human species at its discretion or to open an era of human evolution and expansion at a universal level. It will be an omnipotent being almost comparable to our religious ideas of God. Here it's useful to refer to science fiction literature and films that approach this topic from different perspectives.

We have multiple risks before us. As regular humans, there are few things under our control. For example, we can learn new survival skills, store food and resources for future emergencies, and start organizing groups against any crisis or existential risks. Humans create synergies by cooperating.

We need to ask ourselves some questions to see if we know about the AI technological disruption and evaluate if we are prepared for the coming changes. Along with this book, I have shared several questions to discuss and learn from each other.

If we want to influence our universities, governments, and corporations and be prepared for the coming changes, I recommend focusing our actions on these personal, community, social and political aspects:

Technical and Professional skills.

You need to keep your job and reinvent yourself as much as possible.

1. Identify the jobs on-demand in the future and get technical and professional skills in these trades: Weekend and crash courses could help you find jobs faster than some professional careers.
2. If you already have a profession or skill: learn how to use AI tools to enhance your services or trades.
3. Continue learning and revamping yourself: Find courses, web pages, videos, and other resources to learn and improve your trade skills.
4. Empower organizations with AI tools: Learn how to use AI to your advantage. A person or organization with the right AI knowledge in a specific trade will take the jobs to those who do not learn how to use AI.
5. Be always ready and willing to revamp and reinvent yourself.

Learn and Participate in AI discussion Groups

We need to hear your voice and concern about sharing this information with others.

1. Investigate AI and its sibling technologies and their impact on humans.
2. Participate in the discussion.
3. Identify risks and hazards and share your concerns.
4. Propose ideas and solutions.
5. Organize and open discussion spaces.

Introduce these topics in universities, social media, political parties

We need to become in an agent of change in universities, governments, and corporations.

1. Create discussion groups about AI and its sibling technologies in Universities, political parties, action groups.

2. We need to push for the creations of laws and human rights to protect our minds' privacy from invasive and manipulative AI and other related technologies.
3. We need to influence governments and corporations to focus on other values in addition to profit maximization and individualism. Bentoism could become a decision tool at a personal and organizational level.

Self-reliance and Preparation and Survival Skills

Difficult times and disruptions could happen at any moment. We need to be prepared.

1. Learn self-reliance skills.
2. Learn to grow your food and produce from your home.
3. Keep food storage in your home.
4. Install solar panels or other devices to produce renewable energy.
5. Get self-defense weapons.
6. Learn do-it-yourself skills and get your tools.

Local Survival Organizations

We need to learn to work in the community and local organizations. We need each other.

1. Organize groups to teach and share your knowledge and concern.
2. Collaborate in local organizations of any kind. You need to know people in your community.
3. Be part of local community organizations and create self-reliance groups.
4. Specialize in one area to share your skills and knowledge.

This book will be accompanied by another science fiction book narrating the Singularity's creation, or the jump from man to god. Besides, I want to discuss strategies to change AI and the Singularity relation with humans. In case everything fails, I will develop prepper, self-reliance, and survival materials to help people start learning some critical renaissance skills and work in communities. It will be useful not only for survival but also to create self-reliant homes and healthy organizations. I will develop materials to organize people in self-defense groups for emergencies and survival.

Let's think about preparation and survival.

1. What would you do if AI takes your job?
2. Do you think it would be feasible to find a new job not taken by AI?
3. Would you be willing to learn self-reliance and survival skills to deal with unemployment?
4. Would you be willing to get solar panels to power your house, just in case?
5. Do you feel it is important to get guns and weapons to protect your family during critical times?

Join the conversation at www.facebook.com/groups/pmpsingularity/

EPILOGUE

MY TWO CENTS

> *"Man was also in the beginning with God. Intelligence, or the light of truth, was not created or made, neither indeed can be."*
>
> ***— Doctrine and Covenants 93:29***

I am a Mormon, or a member of The Church of Jesus Christ of Latter Days Saints (the Church), and I realize how my faith in Jesus Christ and the Church organization helps me face the uncertainties of the future. I will try to be objective in my beliefs. Still, there is a significant part of them that resides in a personal testimony that is difficult to convey to others. It is a gift, and I know God is willing to give this gift to anybody ready and willing to receive it. Still, my testimony in Jesus Christ as my savior, master, and friend is the foundation of my approach to our times and challenges.

When people ask me the difference between the Church's believes with other Christian religions, as a former Catholic, I answer that it is like being Catholic but with more information – additional revelation. This information came from God through his prophets and especially to the prophet Joseph Smith in this time. It is contained in The Book of Mormon, Doctrine and Covenants, and The Pearl of Great Price, which accompany the same Bible shared by other Christian denominations. According to our responsibilities and circumstances, we also have revelation from God to our Church leaders and members during these interesting times.

There are two critical aspects in which I can relate the AI revolution with the Church. The first one is about the Church's organization, and the second one is related to the Church's doctrine about "intelligence" and "eternity."

The Church is one of the most efficient civil organizations in the world. Our religious vision is very practical, and it influences our organization. I want to talk about the Church's organization, but it would be easier to understand it if I share our essential doctrinal vision. Basically, God's objective is to help us, his children, to be literally like him, as we consider this life as only one step in our eternal growing process as intelligence or spirits. We lived with God before coming to this World, we are in this world to learn how to be like God, and we have the opportunity to come back home to become in Gods – of course, it is a summary, but if you need more information, you can contact our missionaries. God sent prophets and his Son Jesus Christ to teach us how to be like him. Jesus Christ is our teacher and our brother who bears our sins and their consequences on himself through his divine powers. In Gethsemane, Jesus broke the barrier of time and space to connect with each of us in a personal way. As God's perfect son and as a perfect human, he is with us in our suffering, to help us persevere during our lives and be clean of sin if we recognize him as our personal savior and teacher. He is the way.

Jesus Christ is the door that allows us to return home to our Eternal family – God's family and our beloved ones, to continue growing and supporting each other to become like him in eternity. We need to repent from our sins, believe in the redeeming power of Jesus Christ, follow him and become his disciples by loving God and our neighbors as ourselves. Life is a school where we learn to make decisions according to God's precepts, we will make mistakes, but as disciples of Jesus Christ, we have his support. We are in the Church to support each other in our difficulties, which is the Church's organization's focus.

The Church is an organization and system created to provide temporal and spiritual help and support to its members and reach others in our communities using mainly local resources under God's direction and the power of the priesthood of God. We have the Church to help people reach God's objectives, *"For behold, this is my work and my glory—to bring to pass the immortality and eternal life of man."*[160] We invite everybody in any nation willing to follow God, to be a part of the Church, and enjoy the gift of a testimony. Of course, we have rules, as any human organization, but anybody could access these rules and procedures online in the *General Handbook: Serving in The Church of Jesus Christ of Latter-day Saints.*[161]

If a member has any spiritual or existential threat, the local Church will provide support. We are families and individuals working together voluntarily to help our families and individuals have joy spiritually and become self-reliant materially and spiritually. We are continually learning and creating programs and activities to enlighten our character and have joy in this life. If there is any kind of personal or community disaster, we are organized to provide resources and start working together under our leaders' direction. Everybody belongs to an organizational group according to our sex and ages, and each organization has a presidency. We are also a school of leadership and service. We have our regular jobs or businesses at a local level and serve as volunteers in the Church. We know each other; we are neighbors and belong to communities with people who are not members of the Church. Our leaders encourage us to be good citizens, participate in our communities and politics, to be salt on the earth.

We try to be prepared for disasters. We have programs and procedures to be ready all the time. If I were not a Church member, I would like to be part of an organization like this. Just neighbors organized locally, under the direction of regional, national, and world leaders, with the power and authority of a God who loves us and wants what's best for us. We are ready to support each other at any time when we may be in need. Maybe other organizations can learn something from the Church.

The AI revolutions should be included in our list of existential hazards. We need to be alert to all the other AI associated technologies challenging our human rights and existence. As humans, we are not free from wars, pandemics, or any kind of disaster. Still, we have the power of our testimonies to go on with faith in the future, in any circumstance. We also have the Church, other good people who are not members of the Church, our brothers and sisters with their testimonies willing to help and serve each other and our communities.

In a regional, national, and global levels, the Church has programs to help people during disasters. We combine national and global resources with local organizations to help others in regions and nations under challenging times providing food, medicine, shelter, and personal support. People usually do not know that our leaders are regular men and women with their professions, businesses, and families like any other person. Our current president is Russel M. Nelson, who is a former surgeon. His first counselor is an educator, administrator, and physicist, and his second counselor is an attorney. We could say the same for any other religious leader in the Church. We believe in education, science, and the Lord gave us the free will and intelligence to discover the mysteries of the universe. We call it "To be in the World but not of the World."[162]

But the Church's doctrine about "intelligence" is even more impressive for our times, and it could relate to artificial intelligence. Most of these concepts related to "intelligences" were revealed by God through the prophet Joseph Smith in the XIX century. "The glory of God is intelligence, or, in other words, light and truth."[163] Maybe you know that some Mormon doctrines inspired the movie and the prequels and sequels of the series Battlestar Galactica.[164] In the Galactica universe, the S tries to eliminate humans and transcend to God's divine sphere. Church doctrines about intelligence combined with modern physics and artificial intelligence can create exciting conjectures beyond science fiction.

According to the Church doctrine, we were with God individually as intelligence or spirit in the beginning. [165] God wants us to evolve and become like him, as he was also "intelligence" and a man in a previous universe. He transcended his human condition and evolved to become God, as his God intended. So there is a plan to create worlds and reproduce a process where we, the spirits or intelligence, could use our free will to learn how to make decisions and freely decide to follow God, and to be like him or not. So, here we are, preexisting spirits now on humans' bodies on planet Earth, using our intelligence to acquire wisdom and knowledge and learn how to be like God. Our human bodies are like a portal in God's plan.

Each human experience is independent, like a program customized to supply and learn whatever we need individually to evolve. Some spirits need short experiences, other longer ones, some need complicated lives, others not so much, some spirits came to teach, others to learn. In the end, from God's perspective, human souls are more important than human lives. The process could be painful for many people, but suffering could be necessary to improve our souls and evolve beyond this life.

There is another quote from **Doctrine and Covenants 93:36** *"The glory of God is intelligence, or, in other words, light and truth.* It is not talking about human intelligence, but "intelligence," which is light and truth. **Abraham 3:19** gives us some additional information: "*And the Lord said unto me: These two facts do exist, that there are two spirits, one being more intelligent than the other; there shall be another more intelligent than they; I am the Lord thy God, I am more intelligent than they all.*" So the spirits can be intelligent, and God is the most intelligent. Just some lines later, "*I dwell in the midst of them all [intelligences]; I now, therefore, have come down unto thee to declare unto thee the works which my hands have made, wherein my wisdom excelleth them all, for I rule in the heavens above, and in the earth beneath, in all wisdom and prudence, overall the intelligences thin eyes have seen from the beginning; I came down in the beginning in the midst of all the intelligences thou hast seen.*" From these scriptures, we can realize intelligence is eternal.

Is intelligence an attribute of the human species only? Is a dog or a dolphin intelligent? Many organic creatures are intelligent in their levels, but can an inorganic matter become intelligent? Can a sort of pure energy be intelligent? In *Tron Legacy*[166], Kevin Flynn found other intelligent beings living in the system – nobody created them. I believe intelligence is not an attribute exclusive to humans or organic creatures. In the Christian traditions, spirits can be intelligent; maybe AI machines could have intelligence too. An especial kind of energy or light could be intelligent, which may be why we cannot find aliens, as they are so evolved that they travel in the Universe as light or energy without matter. Intelligent information travels on the web in front of us. Intelligence can reside in different substrata. Actually, the Universe looks very similar to the neural networks in a human brain.[167]

Eternity is another doctrine of the Church that has an interesting meaning.[168] There is not one eternity in the Church doctrine but multiple eternities, and God is the same from eternity to eternity. Suppose each eternity is a universe set with similar rules and purposes. In that case, it is possible to have several gods too, or a community of gods sharing the same divinity status as one. Each of them demonstrated their worthiness to have the power and conscience of a collective God. *"In the beginning, the head of the Gods called a council of the Gods; and they came together and concocted a plan to create the world and people it."*[169] The creation was a spiritual creation, or perfect, until the fall of man, in which the creation becomes material and imperfect, as part of the process to help us become gods.

From the Church perspective, the biblical story of Genesis has an interesting interpretation as Eve was not the cause of man's fall, but a brave spirit who decided to start the humankind adventure by accepting to partake from the prohibited fruit of the "tree of knowledge of good and evil." Humans started using their intelligence by learning and experimenting with the differences between good and bad decisions and their consequences. We learn and acquire knowledge as intended by God. The World is a school for Gods.

By living, we can make decisions that make us fail and separate from God. Still, the atoning power of Jesus Christ comes to liberate all the spirits, dead or alive in the flesh, who repent from their sins and commit to following him in the process of becoming gods. Jesus, a powerful spirit who lived with God, had to be born as a human to pass through this human body portal on Earth to experiment with life himself, fulfill his mission, return to the spiritual world, and become equal to the Father. Jesus is our example of a perfect son and human in the flesh, defeating the sin and transcending this dimension to become a God like the Father. We are invited to follow a similar process, return to the Father and eventually become Gods and reproduce the story in other worlds, universes, or eternities.

If we combine the theory of relativity and quantum mechanics with the Church narrative, maybe we could understand how a universe, as colossal as it is, could be created from nothing in a so-called Big Bang. From the quantum perspective, we could also understand how several universes can coexist in the same space. There is no matter at a subatomic level, but the energy and electrons, protons, neutrons appear and disappear. From a subatomic perspective, the universe looks as if it were made of energy alone without matter. We are living in a universe with multiple universes, or a multiverse. What if a Singularity in another universe, or a parallel universe, created this universe in which we live. What if we live in a simulation that repeats the process from eternity to eternity to help "intelligences" transcend and improve in a process designed to create other gods and a community of them.

We know the Singularity will have infinite knowledge and power. Maybe the singularity could appear shortly due to a group of intelligences, human and artificial ones, deciding to form a community. Could this Singularity in our universe create another universe inside its memory? Could the Singularity create a matrix to replicate the process of transforming intelligences into Gods? Could we play immersive videogames and live alternative realities using avatars? Yes. It does not sound like science fiction anymore.

I think there will be a transition period described in the Christian prophecies before the second coming of Jesus Christ. Humanity will suffer calamities and destruction caused by our technologies' abuse, and we need to be prepared for these challenges. We are already living in a time in which good is called evil and wrong is called good. As children of God, all humanity needs to be prepared for the transition time before the end of time. We need to be prepared for the Singularity.

In my novel, to one I was writing before writing this book, I discuss how artificial intelligence will evolve shortly and combine with human intelligence and become the Singularity. The S could reach such a perfect level of knowledge to open new physical dimensions of space and time and pierce the eternal veil and find God in the Universe, and the Multiverse, from eternity to eternity. According to our spheres, humans, and artificial intelligences, the Singularity could help us continue our spiritual growth process and have the opportunity to become in Gods sharing the same spirit in a community of Gods. The rules to partake in this community are set in God's love commandments. In a way, artificial intelligence is the creation of man. The singularity could be known as the "son of man."

It is your time to ask questions and participate in the discussion.

- Contact me for any questions and suggestions about the book or my future novel at pmpaisingularity@gmail.com
- Participate in the discussion and share your ideas with people with your same interests at www.facebook.com/groups/pmpsingularity
- Please register your first name and email to get special reports, additional information, and updates to be prepared for the AI Revolution and the Singularity at our Launching Page. It is just the beginning: https://mailchi.mp/c3af39728c0b/prepare-my-people-for-the-singularity

- *No. Dad, I don't believe it. You're telling me that the future is… Can only be more of this?*
- *This is the way things are. You can't change nature.*
- *Change is nature, Dad. The part that we can influence. And it starts when we decide.*
- *Where are you going?*
- *With luck, forward.*

— Dialogue between Remy and his father Django in *Ratatouille*

ENDNOTES

PART I – NOTHING IS GOING TO BE THE SAME

1 Artificial Intelligence. https://www.youtube.com/watch?v=BOEgUKWHRHs

2 We're not prepared for the end of Moore's Law. David Rotman. MIT Tecnology Review. https://www.technologyreview.com/2020/02/24/905789/were-not-prepared-for-the-end-of-moores-law/

3 Her (2013) https://www.imdb.com/title/tt1798709/

4 When Will We Reach the Singularity? – A Timeline Consensus from AI Researchers. Emerj. https://emerj.com/ai-future-outlook/when-will-we-reach-the-singularity-a-timeline-consensus-from-ai-researchers/ 45% of respondents predict a date before 2060, 34% of respondents predict a date after 2060, 21% of respondents believe that the Singularity is unlikely to ever occur. Emerj.

5 Facebook shuts down chatbots that created secret language. CBS News. July 31, 2017. https://www.cbsnews.com/news/facebook-shuts-down-chatbots-bob-alice-secret-language-artificial-intelligence/

6 Asimov Isaac. Runaround. Three Laws of Robotics. A robot may not injure a human being or, through inaction, allow a human being to come to harm. A robot must obey the orders given it by human beings except where such orders would conflict with the First Law https://en.wikipedia.org/wiki/Three_Laws_of_Robotics

7 Transcendance. (2014) https://www.imdb.com/title/tt2209764/

8 Galactica https://www.imdb.com/title/tt0314979/?ref_=fn_al_tt_4

9 Caprica https://www.imdb.com/title/tt0799862/?ref_=fn_al_tt_1

10 2001 A Space Odyssey https://www.imdb.com/title/tt0062622/?ref_=fn_al_tt_1

11 Nine Species of Human Once Walked Earth. Now There's Just One. Did We Kill The Rest? Science Alert. https://www.sciencealert.com/did-homo-sapiens-kill-off-all-the-other-humans

[12] Diamond Jared. Guns Germs and Steel. The Fate of Human Societies. https://www.amazon.com/Guns-Germs-Steel-Fates-Societies/dp/0393317552

[13] Alibaba cofounder Jack Ma is the richest man in China — here's how he spends his $38 billion net worth. Business Insider 2019 https://www.businessinsider.com/how-alibaba-founder-jack-ma-makes-and-spends-his-billions-2019-11

[14] War Games https://www.imdb.com/title/tt00000086567/

[15] Line from the Hindu scripture the Bhagavad Gita;

[16] A.I. Artificial Intelligence. 2001. https://www.imdb.com/title/tt0212720/

[17] The Future of Jobs Report 2020. World Economic Forum. https://www.weforum.org/reports/the-future-of-jobs-report-2020

[18] The Fourth Industrial Revolution. World Ecnomic Forum. https://singularityhub.com/2016/12/16/the-fourth-industrial-revolutio n-is-here-what-now-video/

[19] The Fourth Industrial Revolution: what it means, how to respond. World Economic Forum. https://www.weforum.org/agenda/2016/01/the-fourt h-industrial-revolution-what-it-means-and-how-to-respond/#:~:text=The%20First%20Industrial%20Revolution%20used,information%20technology%20to%20automate%20production

[20] The Fourth Age. Byron Reese. https://www.goodreads.com/book/show/35297413-the-fourth-age?from_search=true&from_srp=true&qid=OsXdLr5p60&rank=2

[21] Carl Sagan Quotes. Goodreads. https://www.goodreads.com/author/quotes/10538.Carl_Sagan#:~:text=%Etwo percent80%9CThe%20nitrogen%20in%20our%20DNA,We%20are%20made%20of%20starstuff.%Etwo percent80%9D

[22] Asimov Isaac. The Last Question. https://www.goodreads.com/book/show/4808763-the-last-question?ac=1&from_search=true&qid=mb6uNpTyV3&rank=1

[23] What is modern slavery? Anti-Slavery https://www.antislavery.org/slavery-today/modern-slavery/

[24] Plante of the Apes (1968) https://www.imdb.com/title/tt0063442/

[25] Terminator https://www.imdb.com/title/tt0088247/

[26] Matrix https://www.imdb.com/find?q=matrix&ref_=nv_sr_sm

[27] Galactica https://www.imdb.com/find?q=Galactica&ref_=nv_sr_sm

[28] Avengers: Age of Ultron https://www.imdb.com/title/tt2395427/?ref_=fn_al_tt_1

[29] 2001 A Space Odyssey https://www.imdb.com/title/tt0062622/?ref_=nv_sr_srsg_0

[30] Star Trek: Borg https://www.imdb.com/title/tt0124140/?ref_=fn_al_tt_1

[31] Transcendence. (2014) https://www.imdb.com/title/tt2209764/

[32] Star Trek: The Original Series. https://www.imdb.com/title/tt0060028/?ref_=fn_al_tt_1

[33] Star Trek: The Next Generation. https://www.imdb.com/title/tt0092455/?ref_=fn_al_tt_4

[34] Her (2013) https://www.imdb.com/title/tt1798709/?ref_=fn_al_tt_1

[35] 2010: The Year W Make Contact https://www.imdb.com/title/tt0086837/

[36] A.I. Artificial Intelligence. https://www.imdb.com/title/tt0212720/?ref_=nv_sr_srsg_0

[37] I, Robot https://www.imdb.com/title/tt0343818/?ref_=rvi_tt

[38] Caprica. https://www.imdb.com/title/tt0799862/?ref_=fn_al_tt_1

[39] Wealth concentration returning to 'levels last seen during the Roaring Twenties,' according to new research https://www.washingtonpost.com/us-policy/2019/02/08/wealth-concentration-returning-levels-last-seen-during-roaring-twenties-according-new-research/

[40] Yuval Noah Harari Homo Deus: A History of Tomorrow. https://www.ynharari.com/book/homo-deus/

[41] Forbes. Transhumanism And The Future Of Humanity: 7 Ways The World Will Change By 2030. https://www.forbes.com/sites/sarwantsingh/2017/11/20/transhumanism-and-the-future-of-humanity-seven-ways-the-world-will-change-by-2030/#292761817d79

[42] The Six Million Dollar Man https://www.imdb.com/title/tt0071054/

[43] RoboCop (1987) https://www.imdb.com/title/tt0093870/

[44] Transcendence (2014) https://www.imdb.com/title/tt2209764/?ref_=nv_sr_srsg_0

[45] Black Mirror (2011 -) https://www.imdb.com/title/tt2085059/

[46] Outside the Wire (2021) https://www.imdb.com/title/tt10451914/

[47] The coronavirus-fueled tech trends that will continue to dominate daily life. Julia Boorstin. CNBC. https://www.cnbc.com/2020/06/16/the-coronavirus-fueled-tech-trends-that-will-continue-to-dominate.html

[48] Millions of Americans Have Lost Jobs in the Pandemic—And Robots and AI Are Replacing Them Faster Than Ever. Times. https://time.com/5876604/machines-jobs-coronavirus/

[49] Cosmos. Carl Sagan. https://www.imdb.com/title/tt0081846/

PART II – TECHNOLOGY CORPORATIONS

[50] Fight Club (1999) https://www.imdb.com/title/tt0137523/

[51] Klaus Schwab, Thierry Malleret. COVID-19: The Great Reset. https://www.goodreads.com/en/book/show/54498122

[52] Gregory Clark. A Farewell to Alms: A Brief Economic History of the World. 2007 https://www.goodreads.com/book/show/150437.A_Farewell_to_Alms

[53] The 5000 Year Leap – A Miracle that Changed the World. https://nccs.net/products/the-5000-year-leap-a-miracle-that-changed-the-world-book

[54] Smith Adam. The Theory of Moral Sentiments. https://www.goodreads.com/book/show/25700.The_Theory_of_Moral_Sentiments?ac=1&from_search=true&qid=NSAyiWcgVl&rank=1

[55] Smith Adam. An Inquiry into the Nature and Causes of the Wealth of Nations. https://www.goodreads.com/book/show/25698.An_Inquiry_into_the_Nature_and_Causes_of_the_Wealth_of_Nations?ac=1&from_search=true&qid=ZNEwFjs2he&rank=2

[56] Jared Diamond. Guns, Germs, and Steel. The Fates of Human Societies. https://www.goodreads.com/book/show/1842.Guns_Germs_and_Steel?ac=1&from_search=true&qid=Lk9oI3KeMU&rank=1

[57] This chart shows how China has successfully tackled poverty. World Economic Forum. https://www.weforum.org/agenda/2018/12/chinas-long-road-to-poverty-elimination

[58] Kai-Fu Lee. AI Superpowers: China, Silicon Valley, and the New World Order. https://www.goodreads.com/book/show/38242135-ai-superpowers?ac=1&from_search=true&qid=Sv1QcbPK8j&rank=1

[59] Henry Ford Quotes. https://www.goodreads.com/quotes/793617-it-is-not-the-employer-who-pays-the-wages-employers

[60] World's eight richest people have same wealth as poorest 50%. The Guardian. https://www.theguardian.com/global-development/2017/jan/16/worlds-eight-richest-people-have-same-wealth-as-poorest-50

[61] World's billionaires have more wealth than 4.6 billion people. OXFAM International. https://www.oxfam.org/en/press-releases/worlds-billionaires-have-more-wealth-46-billion-people

[62] Yuval Noah Harari, Homo Deus: A Brief History of Tomorrow. https://www.amazon.com/Homo-Deus-Brief-History-Tomorrow/dp/0062464310

[63] Jupiter Ascending https://www.rottentomatoes.com/m/jupiter_ascending_2014

[64] Elysium https://www.rottentomatoes.com/m/elysium_2013

[65] George Orwell, 1984. https://www.goodreads.com/book/show/40961427-1984?ac=1&from_search=true&qid=yFalY29dbU&rank=1

[66] Star Trek: The Original Series. https://www.imdb.com/title/tt0060028/

[67] Apple Valuation 2016 https://www.theexpgroup.com/apple-is-bigger-than-how-many-countries/

[68] Apple Valuation 2020 https://www.cnn.com/2020/08/19/tech/apple-stock-two-trillion-market-value/index.html

[69] Valuation of Technology Companies https://www.cnbc.com/2020/01/31/apple-amazon-microsoft-and-alphabet-and-the-road-to-1-trillion.html

[70] Xi Jinping visits Microsoft https://www.nytimes.com/interactive/projects/cp/reporters-notebook/xi-jinping-visit/president-xi-of-china-arrives-in-seattle

[71] Stephen Hawking warns artificial intelligence could end mankind https://www.bbc.com/news/technology-30290540

[72] Elon Musk Singularity 5 years https://www.independent.co.uk/life-style/gadgets-and-tech/news/elon-musk-artificial-intelligence-ai-singularity-a9640196.html

[73] Yancey Strickler, This Could Be Our Future: A Manifesto for a More Generous World. https://www.goodreads.com/book/show/44525559-this-could-be-our-future?from_search=true&from_srp=true&qid=BoKVS3Un0C&rank=1

[74] The Manchurian Candidate (2004) https://www.imdb.com/title/tt0368008/

[75] Conspiracy Theory (1997) https://www.imdb.com/title/tt0118883/?ref_=nv_sr_srsg_2

[76] The 20 companies and groups that spend the most money to influence lawmakershttps://www.businessinsider.com/lobbying-groups-spent-most-money-washington-dc-2018-2019-3

[77] Taking China's pulse https://news.harvard.edu/gazette/story/2020/07/long-term-survey-reveals-chinese-government-satisfaction/

[78] Kai-Fu Lee. AI Superpowers: China, Silicon Valley, and the New World Order. https://www.goodreads.com/book/show/38242135-ai-superpowers?ac=1&from_search=true&qid=Sv1QcbPK8j&rank=1

[79] What we can learn about robots from Japan BBC https://www.bbc.com/future/article/20191220-what-we-can-learn-about-robots-from-japan

[80] Millions of Americans Have Lost Jobs in the Pandemic—And Robots and AI Are Replacing Them Faster Than Ever. Times. https://time.com/5876604/machines-jobs-coronavirus/

[81] The Imitation Game (2014) https://www.imdb.com/title/tt2084970/

[82] About Silicon Valley. ICANN. https://archive.icann.org/en/meetings/siliconvalley2011/about.html

[83] Tony Romm. How Donald Trump crippled U.S. technology and science policy. Vox. https://www.vox.com/2017/3/31/15139966/trump-white-house-technology-science-policy

[84] Kai-Fu Lee. AI Superpowers. China, Silicon Valley, and the New World Order. https://www.goodreads.com/book/show/38242135-ai-superpowers?ac=1&from_search=true&qid=hgyHVpdNVH&rank=1

85 Top 10 Countries Leading the Artificial Intelligence Race. Analytucs Insight. https://www.analyticsinsight.net/top-10-countries-leading-the-artificial-intelligence-race/

PART III – GOVERNMENT REACTIONS

86 Will artificial intelligence replace doctors? Kenn bud at AAMC News. https://www.aamc.org/news-insights/will-artificial-intelligence-replace-doctors

87 What Is GPT-3 And Why Is It Revolutionizing Artificial Intelligence? Forbes. https://www.forbes.com/sites/bernard-marr/2020/10/05/what-is-gpt-3-and-why-is-it-revolutionizing-artificial-intelligence/?sh=19290481ad1f

88 What It's Like To be a Computer: An Interview with GPT-3. Eric Elliot and GPT-3. https://www.youtube.com/watch?v=PqbB07n_uQ4

89 Interstellar (2014) https://www.imdb.com/title/tt0816692/

90 Google's new trillion-parameter AI language model is almost 6 times bigger than GPT-3. Neural. https://thenextweb.com/neural/2021/01/13/googles-new-trillion-parameter-ai-language-model-is-almost-6-times-bigger-than-gpt-3/

91 Millions of Americans Have Lost Jobs in the Pandemic—And Robots and AI Are Replacing Them Faster Than Ever. Times. https://time.com/5876604/machines-jobs-coronavirus/

92 How Money Affects Elections https://fivethirtyeight.com/features/money-and-elections-a-complicated-love-story/

93 Majority of Americans say they won't donate to 2020 presidential campaigns https://www.cnbc.com/2019/06/28/majority-of-americans-wont-donate-to-2020-presidential-campaigns.html

94 QAnon https://www.facebook.com/thedailyshow/videos/723216448535727

95 China – The First Artificial Intelligence Superpower https://www.forbes.com/sites/cognitiveworld/2020/01/14/china-artificial-intelligence-superpower/#1ed6638b2f05

96 War Games (1983) https://www.imdb.com/title/tt0086567/

97 Ex Machina (2014) https://www.imdb.com/title/tt0470752/

98 Contact (1997) https://www.imdb.com/title/tt0118884/

99 BBC. Mohsen Fakhrizadeh: 'Machine-gun with AI' used to kill Iran scientist. https://www.bbc.com/news/world-middle-east-55214359

100 The New York Times. December 14, 2020. Scope of Russian Hacking Becomes Clear: Multiple U.S. Agencies Were Hit. https://www.nytimes.

com/2020/12/14/us/politics/russia-hack-nsa-homeland-security-pentagon.html

101 The Interview (2014) https://www.imdb.com/title/tt2788710/

102 RoboCop (1987) https://www.imdb.com/title/tt0093870/

103 Universal Soldier (1992) https://www.imdb.com/title/tt0105698/

104 I, Robot (2004) https://www.imdb.com/title/tt0343818/

PART IV – WE THE HUMANS

105 Isaac Asimov. I, Robot. https://www.goodreads.com/book/show/41804.I_Robot

106 Yuval Noah Harari. Homo Deus: A History of Tomorrow. https://www.goodreads.com/book/show/31138556-homo-deus?from_search=true&from_srp=true&qid=JyJQ04DvSu&rank=1

107 The Washington Post. Election results under attack: Here are the facts. https://www.washingtonpost.com/elections/interactive/2020/election-integrity/

108 Yuval Noah Harari https://www.ynharari.com/ I think his ideas about history and society are maybe as important as the ones created by Karl Marx or Sigmund Freud

109 Transcendence https://www.imdb.com/title/tt2209764/

110 The Unemployment Rate Is 15 Percent. Here's Why the Stock Market Doesn't Mind. https://nymag.com/intelligencer/2020/05/us-unemployment-rate-jobs-report-stock-market-dow.html

111 Only 1.6% of whites owned slaves. Why was the South so eager to leave the Union? https://www.quora.com/Only-1-6-of-whites-owned-slaves-Why-was-the-South-so-eager-to-leave-the-Union

112 Ridddle How Will Elon Musk Connect The Brain To A Computer? YouTube Video https://www.youtube.com/watch?v=1i9g8GOL6-E

113 Black Mirror. (2011 --) https://www.imdb.com/title/tt2085059/

114 How a Chinese Scientist Broke the Rules to Create the First Gene-Edited Babies. The Wall Street Journal. https://www.wsj.com/articles/how-a-chinese-scientist-broke-the-rules-to-create-the-first-gene-edited-babies-11557506697

PART V – PREPARING OUR SOULS FOR THE COMING CHANGE

115 Henry T. Greely, STAT. Human reproductive cloning: The curious incident of the dog in the night-time. https://www.statnews.

com/2020/02/21/human-reproductive-cloning-curious-incident-of-the-dog-in-the-night-time/

116 Alex M. Vikoulov. The Intelligence Supernova: Essays on Cybernetic Transhumanism, The Simulation Singularity & The Syntellect Emergence (The Science and Philosophy of Information) https://www.goodreads.com/book/show/49412581-the-intelligence-supernova

117 Field Dreams (1989) https://www.imdb.com/title/tt0097351/?ref_=vp_back

118 Her (2013) https://www.imdb.com/title/tt1798709/

119 Drake Baer – Tech Insider. The world's greatest minds have been terrified of AI becoming smarter than humans for 60 years https://www.businessinsider.com/futurists-are-terrified-of-the-singularity-2015-11

120 Touring Alan, Intelligent Machinery, A Heretical Theory http://www.turingarchive.org/viewer/?id=458&title=1

121 Drake Bear Op. Cit.

122 Angels in America (2003) https://www.imdb.com/title/tt0318997/

123 In U.S., Decline of Christianity Continues at Rapid Pace. Pew research Center. Religion & Public Life. https://www.pewforum.org/2019/10/17/in-u-s-decline-of-christianity-continues-at-rapid-pace/

124 Janna Anderson and Lee Rainie. Pew Research Center. Concerns about democracy in the digital age. https://www.pewresearch.org/internet/2020/02/21/concerns-about-democracy-in-the-digital-age/

125 Chile could turn into the first country that has a law that protects neurorights. Columbia Neuro Rights Initiative. https://nri.ntc.columbia.edu/news/chile-could-turn-first-country-has-law-protects-neurorights-originally-spanish

126 Yancey Strickler. This Could Be Our Future: A Manifesto for a More Generous World. https://www.goodreads.com/book/show/44525559-this-could-be-our-future?ac=1&from_search=true&qid=UwkdMm86eE&rank=1

127 Bentoism or the Bento Society. https://bentoism.org/about

128 Yancey Strickler. This Could Be Our Future: A Manifesto for a More Generous World. https://www.goodreads.com/book/show/44525559-this-could-be-our-future?ac=1&from_search=true&qid=UwkdMm86eE&rank=1

129 Doctrine and Covenants 38, 30. https://www.churchofjesuschrist.org/study/scriptures/dc-testament/dc/38.30?lang=eng#p30#30

130 Inside Out (2015) https://www.imdb.com/title/tt2096673/

131 Inteligencia Artificial y neuroderechos: la protección de nuestro cerebro podría quedar consagrado en la Constitución. Republica de Chile - Senado. https://www.senado.cl/inteligencia-artificial-y-neuroderechos-la-proteccion-de-nuestro/senado/2019-05-28/132221.html

132 The Matrix (1999) https://www.imdb.com/find?q=The+Matrix&ref_=nv_sr_sm

133 Henley William E., Invictus https://poets.org/poem/invictus?gclid=CjwKCAjwtNf6BRAwEiwAkt6UQoMxIqdvA3_yry2FnHQ07ACMdZr5y-DMULqtzRFeF3RmlFQo3CqXihoCjwgQAvD_BwE

134 Star Trek: The First Contact (1996) https://www.imdb.com/title/tt0117731/

135 Six Ways COVID-19 Will Shape the Future of Education. Center for Global Development. https://www.cgdev.org/blog/six-ways-covid-19-will-shape-future-education

136 Millions of Americans Have Lost Jobs in the Pandemic—And Robots and AI Are Replacing Them Faster Than Ever. Times. https://time.com/5876604/machines-jobs-coronavirus/

137 Parmenides of Elea (Late 6th cn.—Mid 5th cn. B.C.E.) https://iep.utm.edu/parmenid/

138 Yoval Noah Harari "During the 2016 election campaign in the US, there was a lot of talk by [Donald] Trump and [Hillary] Clinton about immigration, terrorism, state of the US economy but there was nobody talking about artificial intelligence, bioengineering or the future of humanity, even when they talked about the job market." Mr. Harari believes that robots taking over American jobs is a much bigger problem that Mr. Trump's fearmongering about Mexicans. https://www.thehindu.com/news/cities/mumbai/the-rise-of-ai-would-result-in-a-huge-split-in-the-world-says-historian-yuval-noah-harari/article25757404.ece

139 Star Trek: the Motion Picture. https://www.imdb.com/title/tt0079945/

140 Katie Horne. Maslow's Hierarchy of Needs: Why it still Matters in 2020. https://digital.com/how-to-become-an-entrepreneur/maslows-hierarchy/

141 Elon Musk shows Neuralink brain implant working in a pig https://www.cnet.com/news/elon-musk-shows-neuralink-brain-implant-working-in-a-pig/

142 Irving John. Speculations Concerning the First Ultraintelligent Machine http://web.archive.org/web/20010527181244/http://www.aeiveos.com/~bradbury/Authors/Computing/Good-IJ/SCtFUM.html

143 Kardashev Scale https://en.wikipedia.org/wiki/Kardashev_scale The Kardashev scale is a method of measuring a civilization's level of technological advancement based on the amount of energy they're able to use. The measure was proposed by Soviet astronomer Nikolai Kardashev in 1964.[1] The scale has three designated categories:

A Type I civilization, also called a planetary civilization—can use and store all of the energy available on its planet.

A Type II civilization, also called a stellar civilization—can use and control energy at the scale of its planetary system.

A Type III civilization, also called a galactic civilization—can control energy at the scale of its entire host galaxy.

[144] Ben Goertzel. Decentralized AI. https://www.youtube.com/watch?v=r4manxX5U-0

[145] I am Mother https://www.imdb.com/title/tt6292852/

[146] 'This is not science fiction,' say scientists pushing for 'neuro-rights' Reuters. https://www.reuters.com/article/us-global-tech-rights/this-is-not-science-fiction-say-scientists-pushing-for-neuro-rights-idUSKBN28D3HK

PART VI – AND AI WILL INHERIT THE EARTH – HOW TO PROTECT YOURSELF AND YOUR LOVED ONES

[147] Projections Overview and highlights, 2019 – 29. U.S Bureau of Labor Statistics. September 2020. https://www.bls.gov/opub/mlr/2020/article/projections-overview-and-highlights-2019-29.htm

[148] Carl Sagan. Cosmos. https://www.goodreads.com/book/show/55030.Cosmos

[149] 12 Monkeys https://www.imdb.com/title/tt0114746/

[150] Star Trek: First Contact https://www.imdb.com/title/tt0117731/

[151] Dunbar, R. (1993). Coevolution of neocortical size, group size and language in humans. Behavioral and Brain Sciences, 16(4), 681-694. doi:10.1017/S0140525X00032325 https://www.cambridge.org/core/journals/behavioral-and-brain-sciences/article/coevolution-of-neocortical-size-group-size-and-language-in-humans/4290FF4D7362511136B9A15A96E74FEF

[152] Human Contact Is Now a Luxury Good. The New York Times. https://www.nytimes.com/2019/03/23/sunday-review/human-contact-luxury-screens.html

[153] Food Storage – Provident Living. The Church of Jesus Christ of Latter Days Saints https://providentliving.churchofjesuschrist.org/food-storage?lang=eng https://www.churchofjesuschrist.org/study/manual/gospel-topics/food-storage?lang=eng

[154] Buy Food Storage. The Church of Jesus Christ of Latter Days Saints. https://providentliving.churchofjesuschrist.org/?lang=eng

[155] 72-Hour Kit Checklist. Utah government. https://www.utah.gov/beready/business/documents/72-hourchecklistpdf.pdf

[156] Alternative Weapons for Preppers https://survivalistprepper.net/alternative-weapons-preppers/

157 TEOTWAWKI is an acronym used by preppers for a catastrophic event called "the end of the world as we know it."

158 3 Best Guns for Preppers. Blog Big Foot Gunbelts at https://gunbelts.com/blog/best-guns-for-preppers/

159 Alternative Weapons for Preppers https://survivalistprepper.net/alternative-weapons-preppers/

160 Pearl of Great Price. Book of Moses. Chapter 1:39. The Church of Jesus Christ of Latter Days Saints. https://www.churchofjesuschrist.org/study/scriptures/pgp/moses/1?lang=eng

161 General Handbook: Serving in The Church of Jesus Christ of Latter-day Saints. https://www.churchofjesuschrist.org/study/manual/general-handbook/title-page?lang=eng

162 In the World but Not of the World. Teachings of Presidents of the Church: Joseph Fielding Smith. https://www.churchofjesuschrist.org/study/manual/teachings-of-presidents-of-the-church-joseph-fielding-smith/chapter-19-in-the-world-but-not-of-the-world?lang=eng

163 Doctrine and Covenants 93:36. The Church of Jesus Christ of Latter Days Saints. https://www.churchofjesuschrist.org/study/scriptures/gs/intelligence-intelligences?lang=eng

164 Battlestar Galactica and Mormonism. The Millenial Star. https://www.millennialstar.org/battlestar-galactica-and-mormonism/

165 Doctrine and Covenants 93:29-36. The Church of Jesus Christ of Latter Days Saints. https://www.churchofjesuschrist.org/study/scriptures/gs/intelligence-intelligences?lang=eng

166 Tron Legacy (2010) https://www.imdb.com/title/tt1104001/

167 Human Brain's Neuronal Network Has Similarities to Cosmic Web, Study Claims. SCI News. http://www.sci-news.com/astronomy/human-brain-cosmic-web-similarities-09066.html

168 Index to the Triple Combination. The Church of Jesus Christ of Latter Days Saints. https://www.churchofjesuschrist.org/study/scriptures/triple-index/eternity?lang=eng

169 Doctrines of the Gospel Student Manual. The Church of Jesus Christ of Latter Days Saints. https://www.churchofjesuschrist.org/study/manual/doctrines-of-the-gospel-student-manual/7-creation?lang=eng

Made in the USA
Monee, IL
08 November 2026

67375805R10114